Applied Mathematical Sciences

Volume 98

Applied Mathematical Sciences

1. *John:* Partial Differential Equations, 4th ed.
2. *Sirovich:* Techniques of Asymptotic Analysis.
3. *Hale:* Theory of Functional Differential Equations, 2nd ed.
4. *Percus:* Combinatorial Methods.
5. *von Mises/Friedrichs:* Fluid Dynamics.
6. *Freiberger/Grenander:* A Short Course in Computational Probability and Statistics.
7. *Pipkin:* Lectures on Viscoelasticity Theory.
8. *Giacoglia:* Perturbation Methods in Non-linear Systems.
9. *Friedrichs:* Spectral Theory of Operators in Hilbert Space.
10. *Stroud:* Numerical Quadrature and Solution of Ordinary Differential Equations.
11. *Wolovich:* Linear Multivariable Systems.
12. *Berkovitz:* Optimal Control Theory.
13. *Bluman/Cole:* Similarity Methods for Differential Equations.
14. *Yoshizawa:* Stability Theory and the Existence of Periodic Solution and Almost Periodic Solutions.
15. *Braun:* Differential Equations and Their Applications, 3rd ed.
16. *Lefschetz:* Applications of Algebraic Topology.
17. *Collatz/Wetterling:* Optimization Problems.
18. *Grenander:* Pattern Synthesis: Lectures in Pattern Theory, Vol. I.
19. *Marsden/McCracken:* Hopf Bifurcation and Its Applications.
20. *Driver:* Ordinary and Delay Differential Equations.
21. *Courant/Friedrichs:* Supersonic Flow and Shock Waves.
22. *Rouche/Habets/Laloy:* Stability Theory by Liapunov's Direct Method.
23. *Lamperti:* Stochastic Processes: A Survey of the Mathematical Theory.
24. *Grenander:* Pattern Analysis: Lectures in Pattern Theory, Vol. II.
25. *Davies:* Integral Transforms and Their Applications, 2nd ed.
26. *Kushner/Clark:* Stochastic Approximation Methods for Constrained and Unconstrained Systems.
27. *de Boor:* A Practical Guide to Splines.
28. *Keilson:* Markov Chain Models—Rarity and Exponentiality.
29. *de Veubeke:* A Course in Elasticity.
30. *Shiatycki:* Geometric Quantization and Quantum Mechanics.
31. *Reid:* Sturmian Theory for Ordinary Differential Equations.
32. *Meis/Markowitz:* Numerical Solution of Partial Differential Equations.
33. *Grenander:* Regular Structures: Lectures in Pattern Theory, Vol. III.
34. *Kevorkian/Cole:* Perturbation Methods in Applied Mathematics.
35. *Carr:* Applications of Centre Manifold Theory.
36. *Bengtsson/Ghil/Källén:* Dynamic Meteorology: Data Assimilation Methods.
37. *Saperstone:* Semidynamical Systems in Infinite Dimensional Spaces.
38. *Lichtenberg/Lieberman:* Regular and Chaotic Dynamics, 2nd ed.
39. *Piccini/Stampacchia/Vidossich:* Ordinary Differential Equations in $\mathbf{R}^n$.
40. *Naylor/Sell:* Linear Operator Theory in Engineering and Science.
41. *Sparrow:* The Lorenz Equations: Bifurcations, Chaos, and Strange Attractors.
42. *Guckenheimer/Holmes:* Nonlinear Oscillations, Dynamical Systems and Bifurcations of Vector Fields.
43. *Ockendon/Taylor:* Inviscid Fluid Flows.
44. *Pazy:* Semigroups of Linear Operators and Applications to Partial Differential Equations.
45. *Glashoff/Gustafson:* Linear Operations and Approximation: An Introduction to the Theoretical Analysis and Numerical Treatment of Semi-Infinite Programs.
46. *Wilcox:* Scattering Theory for Diffraction Gratings.
47. *Hale et al:* An Introduction to Infinite Dimensional Dynamical Systems—Geometric Theory.
48. *Murray:* Asymptotic Analysis.
49. *Ladyzhenskaya:* The Boundary—Value Problems of Mathematical Physics.
50. *Wilcox:* Sound Propagation in Stratified Fluids.
51. *Golubitsky/Schaeffer:* Bifurcation and Groups in Bifurcation Theory, Vol. I.

(continued following index)

C. de Boor K. Höllig S. Riemenschneider

Box Splines

With 57 illustrations

Springer-Verlag
New York Berlin Heidelberg London
Paris Tokyo Hong Kong Barcelona Budapest

Carl de Boor
Center for Mathematical Sciences
University of Wisconsin–Madison
Madison, WI 53705
U.S.A.

Klaus Höllig
Math Institut A
der Universität
Pfaffenwaldring 57
D-7000 Stuttgart 80
Germany

Sherman Riemenschneider
Department of Mathematics
University of Alberta
Edmonton T6G 2H1 Alberta
Canada

Editors

F. John
Courant Institute of
Mathematical Sciences
New York University
New York, NY 10012
U.S.A.

J.E. Marsden
Department of Mathematics
University of California
Berkeley, CA 94720
U.S.A.

L. Sirovich
Division of
Applied Mathematics
Brown University
Providence, RI 02912
U.S.A.

Library of Congress Cataloging-in-Publication Data
De Boor, Carl.
Box splines / C. de Boor, K. Höllig, S. Riemenschneider.
p. cm. – (Applied mathematical sciences ; v. 98)
Includes bibliographical references and index.
ISBN 978-1-4419-2834-4
1. Spline theory. I. Höllig, K. (Klaus) II. Riemenschneider, S. D. III. Title. IV. Series: Applied mathematical sciences (Springer-Verlag New York Inc.) ; v. 98.
QA224.D4 1993
511'.42–dc20 93-5263

Printed on acid-free paper.

Production managed by Ken Dreyhaupt; manufacturing supervised by Genieve Shaw.
Photocomposed copy prepared from the authors' TEX files with reformatting by Springer-Verlag.

9 8 7 6 5 4 3 2 1

Preface

Compactly supported smooth piecewise polynomial functions provide an efficient tool for the approximation of curves and surfaces and other smooth functions of one and several arguments. Since they are locally polynomial, they are easy to evaluate. Since they are smooth, they can be used when smoothness is required, as in the numerical solution of partial differential equations (in the Finite Element method) or the modeling of smooth surfaces (in Computer Aided Geometric Design). Since they are compactly supported, their linear span has the needed flexibility to approximate at all, and the systems to be solved in the construction of approximations are 'banded'.

The construction of compactly supported smooth piecewise polynomials becomes ever more difficult as the dimension, s, of their domain $G \subseteq \mathbb{R}^s$, i.e., the number of arguments, increases. In the univariate case, there is only one kind of cell in any useful partition, namely, an interval, and its boundary consists of two separated points, across which polynomial pieces would have to be matched as one constructs a smooth piecewise polynomial function. This can be done easily, with the only limitation that the number of smoothness conditions across such a breakpoint should not exceed the polynomial degree (since that would force the two joining polynomial pieces to coincide). In particular, on any partition, there are (nontrivial) compactly supported piecewise polynomials of degree $\le k$ and in $C^{(k-1)}$, of which the univariate B-spline is the most useful example. However, when $s > 1$, then a useful partition may contain cells of various types (simplices and perturbations of parallelepipeds being the most common), and the boundary of a cell is not only connected, but becomes an ever more important part of a cell as s increases (since its dimension differs from that of the cell itself only by 1). This makes the construction of low-degree compactly supported smooth piecewise polynomials impossible except on very special partitions.

In fact, for $s > 1$, the only general construction principle presently available is that of the so-called *polyhedral* splines (a.k.a. 'multivariate B-splines'), of which the box spline, the simplex spline, and the cone spline are the most striking examples. These are obtained as the s-dimensional 'shadow' of an n-dimensional polytope (e.g., the standard n-cube, the standard n-simplex, or the positive n-orthant), are piecewise polynomial of degree $\leq n - s$, with support equal to the corresponding projection of the polytope into $\mathbb{R}^s$, and are in $C^{(n-s-1)}$ if the projector used is generic. In any case, since their partition is determined by just how the n-dimensional polytope is projected into $\mathbb{R}^s$, it is usually not possible to prescribe the partition (in line with the fact that, for a generic partition, there are no compactly supported piecewise polynomials of degree k in $C^{(\rho)}$ for ρ close to k). However, it is possible to refine any triangulation to a partition for which sufficiently many (translated) smooth *simplex* splines are available to span a piecewise polynomial space of good approximation power. Alternatively, if the given partition is sufficiently uniform, then there are *box* splines available whose integer translates span a space of piecewise polynomials with that partition and of good approximation power. Such spaces are the multivariate equivalent of the univariate cardinal spline studied intensively by I. J. Schoenberg and others.

As with Schoenberg's cardinal splines, box splines give rise to an intriguing and beautiful mathematical theory, much more intricate and rich, hence less complete at present. The basic facts, however, have been available for several years, albeit in various papers only. Several of these papers are quite long, and are more devoted to the publication of specific new results than to a careful exposition of the theory. We wrote this book to remedy this.

We have not merely organized the available material in some cohesive way, but have also looked quite carefully at the available arguments and, in many cases, modified them considerably, in line with our goal to provide simple and complete proofs.

While we have endeavored to provide an up-to-date bibliography of papers concerned with box splines, we have made no attempt to report here anything more than what we consider to be the basic box spline theory. In particular, we have included nothing about spaces generated by several box splines, since their theory is far from complete at present. Neither have we dealt with the promising theory of exponential box splines.

The book is organized in the following way. In chapter I, we give the various equivalent definitions of a box spline, and derive its basic properties.

We urge readers unfamiliar with box splines to read ahead to the various detailed (bivariate) examples (and construct others of their own). The rest of the book is concerned with various aspects of the principal shift-invariant space generated by a box spline (a.k.a. a cardinal spline space). For this reason, only box splines with integer directions are considered after chapter I. The linear algebra of a cardinal spline space is the topic of chapter II. It highlights the results of Dahmen and Micchelli on linear independence and the kernels of certain related differential and difference operators. Chapter III brings the basic results on approximation order from a box spline space, and includes a discussion of the construction of quasi-interpolants which realize this order. In chapter IV, Schoenberg's beautiful theory of cardinal spline interpolation is discussed in the setting of box splines where it becomes necessary to devote much more effort to the singular case than in the univariate setting. Chapter V begins with a discussion of the convergence of cardinal splines as their degree tends to infinity and continues with the natural relation of cardinal splines to the multivariate Whittaker cardinal series and wavelets. The theory of discrete box splines is developed in chapter VI in close analogy to that of the (continuous) box spline. It provides the basis for the discussion of subdivision algorithms for the generation of box splines surfaces, in the final chapter.

We have left the attribution of results to the notes at the end of each chapter.

We are indebted to colleagues and students for constructive criticism of various drafts. In particular, we thank Rong-Qing Jia and Amos Ron for their help, as well as Alfred Cavaretta and his students, and Kirk Haller, Tom Hogan, Mike Johnson, Scott Kersey, Jörg Peters, Zuowei Shen, Sivakumar, Shuzhan Xu, and Kang Zhao. We also gratefully acknowledge support from the Army Research Office, from the National Science Foundation, and from the National Sciences and Engineering Research Council of Canada. A NATO Collaborative Research Grant made some face-to-face meetings of the authors possible.

Finally, we record here our heartfelt thanks to D. Knuth for TEX, Cleve Moler and Jack Little for PC(and Pro)-Matlab, and the people at Adobe Systems for PostScript, for these wonderful tools allowed us, for better or for worse, to typeset the text and draw the figures exactly as we wanted them to be.

C. de Boor, Madison
K. Höllig, Stuttgart
S. Riemenschneider, Edmonton

Contents

Chapter II · The linear algebra of box spline spaces

Chapter III · Quasi-interpolants & approximation power

Chapter IV · Cardinal interpolation & difference equations

Chapter V · Approximation by cardinal splines & wavelets

Chapter VI · Discrete box splines & linear diophantine equations

Chapter VII · Subdivision algorithms

Notation

In a subject of this complexity, it is hard to avoid a certain notational complexity. We have tried to deal with this difficulty by using the principle that notation should not be more complicated than the mathematical idea it is meant to represent, and by using **default notation**. This means that we use as few symbols as possible, often relying on context and on the particular combination of symbols used to complete the description. While we realize that the resulting lack of redundancy can be difficult, we feel that the alternative of a symbol thicket or forest is even more disheartening. As a simple example, we use the symbol 0 to denote the number zero, as well as the vector or the function 0, in particular the zero matrix or zero linear map. As a more complicated example, consider the forward difference $\Delta_\nu f$ of the function f: if ν is a positive integer and f is a function on $\mathbb{R}$, then $\Delta_\nu f$ is the function $f(\cdot+\nu)-f$, while it is the function $f(\cdot+\mathbf{i}_\nu)-f$, in case f has (at least) ν arguments, and makes no sense if f has more than 1 but fewer than ν arguments. Finally, if ν is an element of $\mathbb{R}^s$, and f is defined on $\mathbb{R}^s$, then $\Delta_\nu f = f(\cdot+\nu)-f$. We have taken care to avoid contradictory definitions when the situation fits more than one of the contexts considered (check, e.g., the example when f is defined on $\mathbb{R}$ and $\nu = 1$).

Here are specific notations and abbreviations used throughout the book.

$\mathbb{N}$, $\mathbb{Z}$, $\mathbb{R}$, $\mathbb{C}$ denote the collection of natural, whole, real, and complex numbers, respectively, with $\mathbb{N} \neq \mathbb{Z}_+$, the set of nonnegative integers. $\mathbb{R}_+$ denotes the nonnegative reals. We write $\#A$ for the cardinality of the set A.

We describe a map $f : X \to Y : x \mapsto f(x)$ in terms of its **domain**

X and its **target** Y, with $\mapsto$ indicating the particular rule by which the value $f(x)$ of f at its argument x is to be obtained. We denote by $\operatorname{ran} f := f(X) := \{f(x) : x \in X\}$ the **range** of such f. We use a small circle to denote map composition; thus, if also $g : Y \to Z$, then

$$g \circ f : X \to Z : x \mapsto g(f(x)).$$

We denote by Y^X the collection of all maps from X to Y, but write Y^n instead of $Y^{\{1,\ldots,n\}}$ and $Y^{m\times n}$ instead of $Y^{\{1,\ldots,m\}\times\{1,\ldots,n\}}$. We call the elements of Y^n n-vectors or n-sequences, and the elements of $Y^{m\times n}$ $m \times n$-matrices, but we do not think of n-vectors as being column matrices or row matrices. Rather, we think of $f \in Y^n$ as an ordered list of n elements from Y. In particular, we feel free to concatenate such lists: if $f \in Y^n$ and $g \in Y^m$, then $h := (f, g)$ is the element of Y^{n+m} which satisfies

$$h(i) = \begin{cases} f(i), & \text{if } i \le n \\ g(i-n), & \text{if } i > n \end{cases}.$$

We use $f(i)$ to denote the ith entry of $f \in Y^n$. Thus, $\xi(2)$ is the second component of the vector $\xi \in \mathbb{R}^n$.

We use concatenation to denote the scalar product of two real vectors. Thus

$$\xi\eta := \xi(1)\eta(1) + \xi(2)\eta(2) + \cdots \quad .$$

We follow MATLAB in the notation for matrices constructed from parts of a given matrix. For example, if $A \in \mathbb{F}^{m\times n}$, P is a p-sequence with values in $\{1, \ldots, m\}$, and Q is a q-sequence with values in $\{1, \ldots, n\}$, then

$$A(P, Q)$$

is the $p \times q$-matrix whose (i, j)-entry is $A(P(i), Q(j))$. Further, it is convenient to use

$$u{:}v$$

for the sequence $(u, u + 1, u + 2, \ldots, v)$. Finally, in this context, : by itself stands for the sequence of all relevant indices. For example, if A is a $m \times n$-matrix and Q is a q-sequence with values in $\{1, \ldots, n\}$, then

$$A(:, Q)$$

is the $m \times q$-matrix whose (i, j)-entry is $A(i, Q(j))$. In particular, we use $A(i, :)$ and $A(:, j)$ to denote the ith row, respectively the jth column, of the matrix A.

We use brackets to indicate matrices in terms of its constituents. E.g., [1] is the 1×1-matrix whose sole entry is 1, while $[A, B]$ ($[A; B]$) is the matrix made up of the columns (rows) of A, followed by the columns (rows) of B. In particular, $[f_1, f_2, \ldots, f_n]$ is the $m \times n$-matrix with *columns* $f_1, f_2, \ldots, f_n$ in case $f_1, f_2, \ldots, f_n$ are elements of $\mathbb{C}^m$. If, more generally, $f_1, f_2, \ldots, f_n$ are elements of some linear space Y, then

$$[f_1, f_2, \ldots, f_n] : \mathbb{C}^n \to Y : c \mapsto \sum f_j c(j)$$

is the linear map from $\mathbb{C}^n$ to Y which carries $c \in \mathbb{C}^n$ to $\sum f_j c(j) \in Y$. For this reason, we often write scalars to the right of vectors (rather than the left).

The transpose of a matrix A is denoted by A^T and the inverse (if any) of A^T by A^{-T}.

We denote the identity matrix by $\mathbb{1}$, with its order always clear from the context, and denote its jth column, i.e., the jth unit vector, by $\mathbf{i}_j$.

If x is an n-vector, then

$$\operatorname{diag} x$$

is the diagonal $n \times n$-matrix whose (i,i)-entry is $x(i)$, all i.

For $f : X \to \mathbb{R}$, we denote by

$$\operatorname{argmax} f := \{a \in X : \forall\{x \in X\}\ f(x) \le f(a)\}$$

the collection of points (if any) at which f takes on its maximum value over X. The argmin is defined analogously.

Here is our notation for certain specific maps: We use χ_I to denote the characteristic function of the set I. We denote by sinc the *sinus cardinalis*, i.e., the map

$$\operatorname{sinc} : \mathbb{R} \to \mathbb{R} : t \mapsto \sin(t/2)/(t/2).$$

For the **α-power function** $x \mapsto x^\alpha := \prod_{\nu=1}^{s} x(\nu)^{\alpha(\nu)}$, we use the abbreviation $()^\alpha$, and the abbreviation $[\![]\!]^\alpha := ()^\alpha/\alpha!$ for the **normalized α-power function**, i.e.,

$$(1) \qquad [\![x]\!]^\alpha := x^\alpha/\alpha! := \prod_{\nu=1}^{s} \frac{x(\nu)^{\alpha(\nu)}}{\alpha(\nu)!}$$

with $\alpha \in \mathbb{Z}^s$ (and $[\![]\!]^\alpha = 0$ if some entry of α is negative). In TEX, we describe $[\![x]\!]^\alpha$ by `\braket{x}^\alpha`, using the definition
`\def\braket#1{\hbox{$\[\![$}#1\hbox{$\]\!]$}}`

We use the abbreviation $|\alpha| := \|\alpha\|_1$, and write $\beta \le \alpha$ ($\beta < \alpha$) in case that inequality holds for every component (and, in addition, $\beta \ne \alpha$).

By $\Pi = \Pi(\mathbb{R}^s)$ we denote all polynomials on $\mathbb{R}^s$, by Π_α we denote the span of all $[\![]\!]^\beta$ with $\beta \le \alpha$, and by Π_q the span of all $[\![]\!]^\beta$ with $|\beta| \le q$. The notations $\Pi_{<\alpha}$ and $\Pi_{<q}$ are defined analogously. When we want to stress that we are dealing with polynomials on some (possibly shifted) plane H in $\mathbb{R}^s$, we write $\Pi(H)$.

We use σ_h to denote the scale map, i.e., $\sigma_h f : x \mapsto f(x/h)$, and use τ_y to denote the translation map, i.e., $\tau_y f : x \mapsto f(x+y)$. We refer to translation by an *integer* vector as a **shift**. The general (constant-coefficient) **difference operator** is of the form $r(\tau) := \sum_{j\in\mathbb{Z}^s} a(j)\tau_j$ for some Laurent polynomial $r := \sum_{j\in\mathbb{Z}^s} a(j)()^j$.

Where appropriate, we write difference operators in the form $p(\Delta)$ or $p(\nabla)$, with $p \in \Pi$, $\Delta_y := \tau_y - 1$ and $\nabla_y := 1 - \tau_{-y}$, and, e.g., $\Delta^j := \prod_{\nu=1}^s \Delta_{\mathbf{i}_\nu}^{j(\nu)}$. Correspondingly, we use D_y to denote differentiation in the direction y, and $D_\nu := D_{\mathbf{i}_\nu}$ for differentiation with respect to the νth argument, and write the (constant-coefficient) **differential operator** in the form $p(D) := \sum_{j\in\mathbb{Z}_+^s} D^j p(0)/j!\, D^j$. We also use

$$D_Y := \prod_{y\in Y} D_y$$

and the analogous abbreviations Δ_Y, ∇_Y, with Y any finite sequence in $\mathbb{R}^s$.

We use $\mathbf{L}_p(G)$ to denote the space of p-integrable functions on G, and $\ell_p(\mathbb{Z}^s)$ for the normed linear space of p-summable mesh-functions, and denote their respective norms by $\| \|_p$. Since we rely heavily on the comma as the separator for lists, we use the double dot .. in the specification of an interval. Thus, $(a\,..\,b)$ is the open interval with endpoints a and b, as distinct from the point (a,b) in the plane. As another illustration, if $a, b \in \mathbb{R}^m$, then

$$[a\,..\,b] := \{ta + (1-t)b : t \in [0\,..\,1]\}$$

is the segment in $\mathbb{R}^m$ with endpoints a and b, as distinct from the $m \times 2$-matrix $[a, b]$ with columns a and b. In TEX, we describe $(a\,..\,b)$ by `(a\fromto b)`, using the definition
`\def\fromto{\mathinner{\ldotp\ldotp}}`

We use the special symbols

$$\blacksquare := [0\,..\,1]^n, \quad \square := [0\,..\,1)^n$$

for the closed, respectively half-open, unit cube, with its dimension, n, understood from the context. For example, if Ξ is a matrix, then $\Xi\blacksquare = \sum_j \Xi(:,j)[0\,..\,1]$ is the zonotope spanned by the columns of Ξ. Since these symbols are non-standard, here are their TEX descriptions (together with versions of use in subscripts):

```
\def\makehsquare#1#2#3{
\dimen0=#1\advance\dimen0 by -#3
\vrule height#1 width#2 depth0pt \kern-#2
\vrule height#1 width#1 depth-\dimen0 \kern-#1
\vrule height#2 width#1 depth0pt \kern-#3
\vrule height#1 width#3 depth0pt
}
\def\boxx{\mathop{\makehsquare{6pt}{1.2pt}{1.2pt}}}
\def\boxo{\mathop{\makehsquare{6pt}{1.2pt}{.05pt}\kern.05em}}
\def\subboxx{\mathop{\makehsquare{4pt}{1pt}{1pt}}}
\def\subboxo{\mathop{\makehsquare{4pt}{1pt}{.05pt}}}
```

We use δ to indicate point-evaluation. Thus $\delta_x f := f(x)$, for whatever x in the domain of whatever function or map f. By extension, we use $\delta = \delta_0$ also to denote the **Kronecker sequence**, i.e., the sequence

$$\delta_0 : \mathbb{Z}^s \to \{0,1\} : \alpha \mapsto \begin{cases} 1, & \alpha = 0; \\ 0, & \text{otherwise.} \end{cases}$$

For, $\delta_0 f = \sum_\alpha \delta_0(\alpha) f(\alpha) = f(0)$ for any f on $\mathbb{Z}^s$.

We often write const to denote a constant, with the precise value of such a constant usually different from one appearance to the next.

In order to make it easy to find items, we number all items, be they equations, theorems, propositions, remarks, examples, etc., consecutively, starting with item 1 in each chapter. Item j in chapter i is referred to by (j) within chapter i, but by (i.j) outside chapter i.

I

Box splines defined

In this chapter, we give various equivalent definitions of the box spline, and derive its basic properties, supplementing the mathematical discussion with several detailed examples and illustrations.

The analytic definition. The **box spline** M_Ξ associated with the $s \times n$ matrix Ξ with columns in $\mathbb{R}^s \backslash 0$ is, by definition, the distribution given by the rule

$$(1) \qquad M_\Xi : C(\mathbb{R}^s) \to \mathbb{R} : \varphi \mapsto \langle M_\Xi, \varphi \rangle := \int_\square \varphi(\Xi t) dt,$$

with $\square := [0\,.\,.\,1)^n$, the half-open unit n-cube.

To illustrate this definition, consider the example $\Xi = [1\ 1]$. In this case, the right hand side of (1) equals

$$\int_{[0..1)^2} \varphi(t_1 + t_2) dt_1 dt_2.$$

As indicated in (2)Figure below, this can be written as $\int M_{[1\ 1]}(x)\varphi(x)dx$, with

$$M_{[1\ 1]}(x) = \begin{cases} x, & \text{if } 0 \le x \le 1; \\ 2 - x, & \text{if } 1 \le x \le 2; \\ 0, & \text{otherwise.} \end{cases}$$

This illustrates that the box spline can be identified with a function (on $\operatorname{ran} \Xi$), as we will show in a moment.

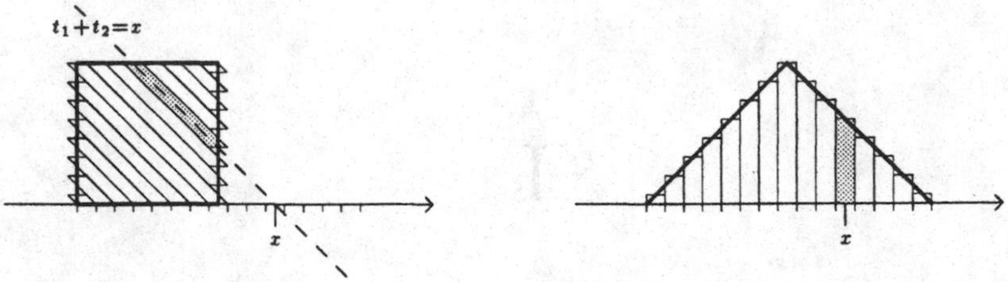

(2)Figure. Summing $(t_1, t_2) \mapsto \varphi(t_1 + t_2)$ over the square $[0\,..\,1]^2$ is the same as summing $x \mapsto M_{[1\ 1]}(x)\varphi(x)$ over the interval $[0\,..\,2]$, since corresponding strips (e.g., the two shaded strips) have equal area (i.e., have the same base and height).

A geometric description. It is also possible to describe M_Ξ in quite geometric terms. For this, we split the integration over $\square$ into two stages. We first integrate in directions along which Ξt, hence $\varphi(\Xi t)$, is constant and after that in directions perpendicular to those (see (10)Figure). In other words, we decompose $\mathbb{R}^n$ into the direct sum $(\ker \Xi) \oplus \big((\ker \Xi)\perp\big)$. From (1) and Fubini's Theorem,

$$\langle M_\Xi, \varphi\rangle = \int_{y \perp \ker \Xi} \int_{z \in \ker \Xi} \chi_\square(y+z)\varphi(\Xi(y+z))\,dz\,dy,$$

with $\chi_\square$ the characteristic function of $\square$. The inner integral is over the kernel or nullspace $\ker \Xi$ of Ξ, i.e., $z \in \ker \Xi$, hence equals $\varphi(\Xi y)\,\mathrm{vol}_{n-d}\big((y + \ker \Xi) \cap \square\big)$, with

$$d := d(\Xi) = \dim \operatorname{ran} \Xi$$

the dimension of the range of Ξ. Now set $x := \Xi y$. Then $y + \ker \Xi = \Xi^{-1}\{\Xi y + 0\} = \Xi^{-1}\{x\}$, and $dx = |\det \Xi| dy$, with $\det \Xi$ the (d-dimensional signed) volume of the image under Ξ of a unit volume in $(\ker \Xi)\perp = \operatorname{ran} \Xi^T$ (necessarily nonzero since Ξ is one-one on $(\ker \Xi)\perp$). Consequently,

$$\langle M_\Xi, \varphi\rangle = \int_{\operatorname{ran} \Xi} \varphi(x)\mathrm{vol}_{n-d}\Big(\Xi^{-1}\{x\} \cap \square\Big)dx/|\det \Xi|,$$

which shows that M_Ξ can be identified with the function

$$(3) \qquad M_\Xi(x) = \mathrm{vol}_{n-d}\Big(\Xi^{-1}\{x\} \cap \square\Big)/|\det \Xi|, \qquad x \in \operatorname{ran} \Xi.$$

In particular, if $n \ge s$ and Ξ has full rank ($s = d$), then M_Ξ is a function on $\mathbb{R}^s$.

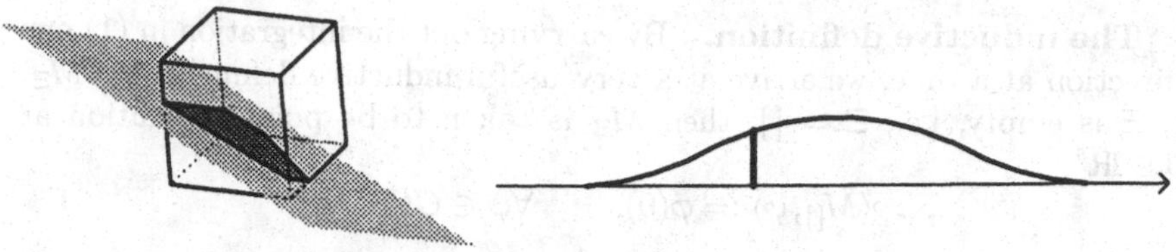

(4)Figure. Geometric definition of M_Ξ illustrated, with $\Xi = [1\ 2\ 1]$ and $x = 4/3$.

Note that, since $\langle M_\Xi, 1\rangle = \int_\square 1dt = 1$, we have

(5) $$\int_{\mathrm{ran}\,\Xi} M_\Xi = 1$$

for the function M_Ξ.

One would compute $|\det \Xi|$ as follows: Let $\Xi = VXU^T$, with the columns of $V \in \mathbb{R}^{s\times d}$ an orthonormal basis for $\mathrm{ran}\,\Xi$ and the columns of $U \in \mathbb{R}^{n\times d}$ an orthonormal basis for $\mathrm{ran}\,\Xi^T = \ker \Xi\perp$. In other words, X is the matrix representation for $\Xi_{|\,\mathrm{ran}\,\Xi^T}$ with respect to orthonormal bases. E.g., VXU^T could be chosen as the singular value decomposition for Ξ. More simply, Gram-Schmidt could be used on the rows of Ξ to produce the factorization $\Xi = WU^T$ which is satisfactory in case V is square. If it is not, i.e., if $d < s$, then Gram-Schmidt applied to the columns of W would give the desired $W = VX$. In any case, the set $U\square$, i.e., the (half-open) parallelepiped **spanned** by the columns of U, has unit volume, hence $|\det \Xi|$ is the volume of the parallelepiped spanned by the columns of $\Xi U = VXU^TU = VX$, therefore the volume of the parallelepiped spanned by the columns of X, i.e., $|\det \Xi| = |\det X|$. In particular, if Ξ has full rank, i.e., if $d = s$, then $\Xi\Xi^T = VXX^TV^T$ with V unitary. Therefore, in this case, $\det \Xi$ can be computed more simply as

(6) $$|\det \Xi| = \sqrt{\det(\Xi\Xi^T)}.$$

Finally, construction of the set $\Xi^{-1}\{x\} = y + \ker \Xi$ requires that one find (a basis for) $\ker \Xi$ and a point $y \in \Xi^{-1}\{x\}$. It seems natural to choose $y \in (\ker \Xi)\perp = \mathrm{ran}\,\Xi^T$. This means that

(7) $$\Xi^{-1}\{x\} = \Xi^T(\Xi\Xi^T)^{-1}\{x\} + \ker \Xi$$

in case $\mathrm{rank}\,\Xi = s$. More generally, in terms of the factorization $\Xi = VXU^T$ mentioned earlier, and with $[U, U_1]$ an orthonormal basis for $\mathbb{R}^n$,

$$\Xi^{-1}\{x\} = UX^{-1}V^Tx + \mathrm{ran}\,U_1.$$

The inductive definition. By carrying out the integration in (1) one direction at a time, we arrive at a very useful inductive definition for M_Ξ. If Ξ is empty, i.e., $\Xi = []$, then M_Ξ is taken to be point evaluation at $0 \in \mathbb{R}^s$,

$$\langle M_{[]}, \varphi \rangle := \varphi(0), \qquad \forall \varphi \in C(\mathbb{R}^s).$$

If $\Xi \cup \zeta$ is any matrix formed from Ξ by the addition of the column $\zeta \in \mathbb{R}^s$, then the box spline $M_{\Xi\cup\zeta}$ is given by the convolution equation

$$M_{\Xi\cup\zeta} = \int_0^1 M_\Xi(\cdot - t\zeta)dt \tag{8}$$

defined by the action

$$\langle \int_0^1 M_\Xi(\cdot - t\zeta)\,dt, \varphi \rangle := \langle M_\Xi, \int_{[0..1)} \varphi(\cdot + t\zeta)\,dt \rangle, \qquad \forall \varphi \in C(\mathbb{R}^s).$$

Definition (1) follows from this last expression using induction.

When Ξ is invertible, (i.e., $d = s = n$), then a change of variables in (1) shows directly that M_Ξ is the normalized characteristic function of the parallelepiped $\Xi\square$; specifically,

$$M_\Xi = \frac{1}{|\det \Xi|}\chi_{\Xi\square}. \tag{9}$$

This equation suggests a convenient starting point for the inductive construction of M_Ξ via (8), i.e., via repeated averaging: Start with $\chi_{Z\square}/|\det Z|$ for some submatrix Z of Ξ whose columns form a basis for ran Ξ, since this is a simple *function* on $\mathbb{R}^s$.

Example. Let $\Xi = [2\ 1]$, hence $n = 2, s = 1$. Then $\ker \Xi$ is spanned by the vector $(1, -2)$ and $(\ker \Xi)\perp$ is spanned by the vector $y := (2, 1)$. (10)Figure illustrates that

$$v(t) := \mathrm{vol}_{2-1}\Big((yt + \mathrm{span}(1, -2)) \cap [0\,..\,1)^2\Big)$$

rises linearly from 0 to $\sqrt{(1/2)^2 + 1^2} = \sqrt{5}/2$ on $[0\,..\,\alpha]$, with α such that $(2, 1)\alpha + (1, -2)\beta = (0, 1)$, i.e., $\alpha = 1/5$. Further, $v(t)$ is constant on $[\alpha\,..\,2\alpha]$, and falls linearly from $\sqrt{5}/2$ to 0 on $[2\alpha\,..\,3\alpha]$, and is zero otherwise.

Since $\Xi\Xi^T = [5]$, we compute $|\det \Xi| = \sqrt{5}$ from (6), and, for $x \in \operatorname{ran}\Xi = \mathrm{span}(\mathrm{i}_1) = \mathbb{R}^1$, $\Xi^{-1}\{x\} = y[1/5]x + \mathrm{span}(1, -2)$ from (7). We conclude that

$$M_{[2\ 1]}(x) = v(x/5)/\sqrt{5},$$

i.e., $M_{[2\ 1]}(x)$ rises linearly from 0 to 1/2 on $[0\,..\,1]$, takes the value 1/2 on all of $[1\,..\,2]$, and falls linearly from 1/2 to 0 on $[2\,..\,3]$. In particular, $\int M_{[2\ 1]} = 1/4 + 1/2 + 1/4 = 1$, as it should be, by (5).

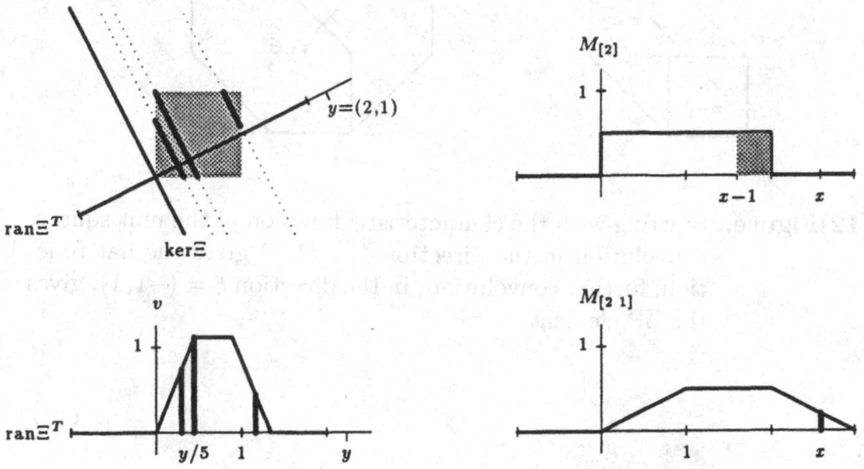

(10)Figure. $M_{[2\ 1]}$ constructed from sections and from averaging.

For the inductive definition, we could start with $M_{[2]}$ which, according to (9), equals $\chi_{[0..2)}/2$, and so find from (8) that

$$M_\Xi = \int_0^1 \chi_{[0..2)}(\cdot - t)\, dt/2,$$

which can be seen to give the same piecewise linear function obtained earlier; see (10)Figure. □

(11)Example. As a first bivariate example, we consider the inductive construction of M_Ξ for $\Xi = \begin{bmatrix} 1 & 0 & 1 & -1 \\ 0 & 1 & 1 & 1 \end{bmatrix}$. Here, we start with the characteristic function M_Z of the unit square, i.e., with $Z = \begin{bmatrix} 1 & 0 \\ 0 & 1 \end{bmatrix}$. We average this function in the direction $\zeta := (1, 1)$ and so obtain the **hat function** or **Courant element**. In the next and final step, we average the hat function in the direction $\xi := (-1, 1)$ and obtain the **Zwart-Powell element**, or **ZP element**, for short. In (46)Example, we present four different ways of computing this element. □

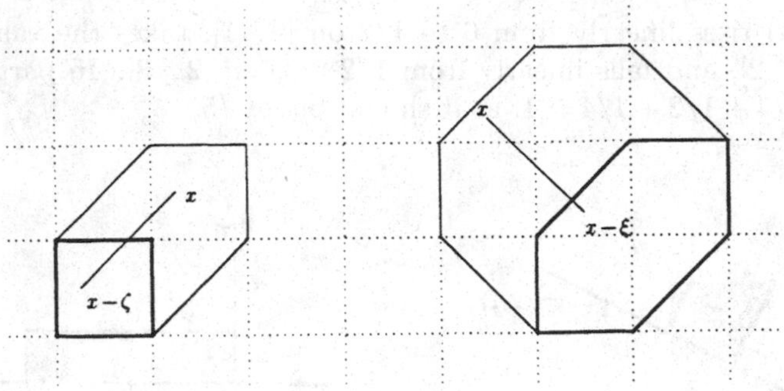

(12) Figure. Starting with the characteristic function of the unit square, convolution in the direction $\zeta = (1,1)$ gives the hat function; further convolution, in the direction $\xi = (-1,1)$, gives the ZP element.

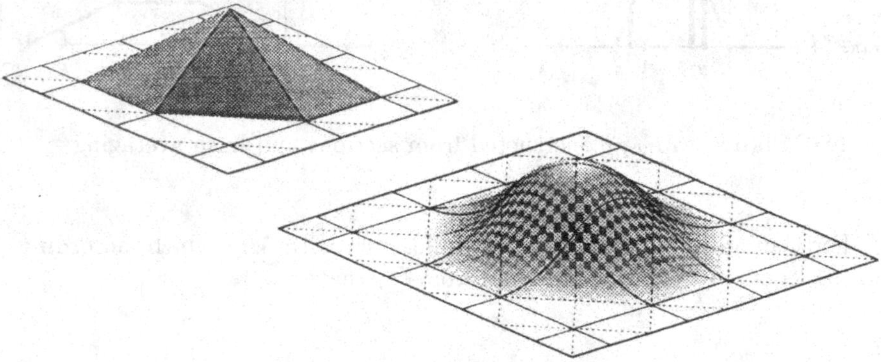

(13) Figure. The hat function and the ZP element.

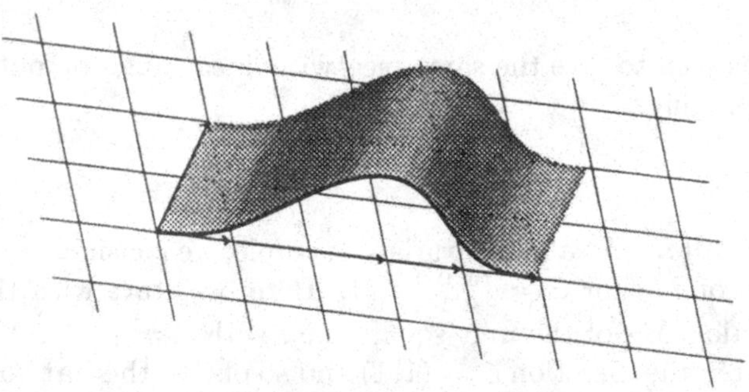

(14) Figure. A box spline constant in one direction ($\Xi = \begin{bmatrix} 1 & 2 & 1 & 1 & 1 \\ 0 & 0 & 0 & 0 & 2 \end{bmatrix}$).

The inductive definition shows that the box spline $M_{\Xi\cup\zeta}$ is piecewise constant in the direction ζ in case $\zeta \notin \operatorname{ran}\Xi$. This is illustrated in (14)Figure.

The inductive definition also shows that M_Ξ is piecewise polynomial (on $\operatorname{ran}\Xi$), with each (nontrivial) polynomial piece of degree at most

$$k(\Xi) := \dim\ker\Xi.$$

More precise statements are given in (28) and (37).

Finally, here are two more sophisticated box-splines.

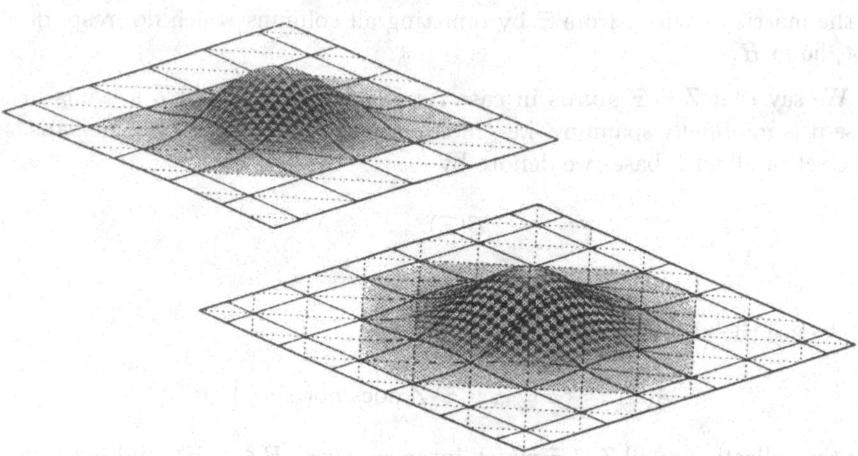

(15)Figure. M_Ξ for $\Xi = \begin{bmatrix} 1 & 1 & 0 & 0 & 0 & 1 \\ 0 & 0 & 1 & 1 & 1 & 1 \end{bmatrix}$ and $\Xi = \begin{bmatrix} 1 & 0 & 1 & 1 & -1 & -1 \\ 0 & 1 & 1 & 1 & 1 & 1 \end{bmatrix}$.

Conventions. The following conventions involving the matrix Ξ and its submatrices will be used throughout.

M_Ξ does not depend on the order in which the columns making up the matrix Ξ appear in Ξ, but it does depend on the possible multiplicities with which any particular vector appears as a column of Ξ and not just on the set of vectors making up the columns of Ξ. Correspondingly, we write

$$n =: \#\Xi$$

for the number of columns of Ξ, and we intend to run through all the columns ξ of Ξ counting multiplicities when we write $\xi \in \Xi$.

More generally, we use the notation

$$\mathrm{Z} \subseteq \Xi$$

to indicate that we are running through all matrices Z obtained from Ξ by the deletion of columns. Even though two such matrices might be the same as matrices, we think of them as different unless they are obtained from Ξ by the omission of the same columns. Similarly, $\Xi \cup \mathrm{Z} := [\Xi, \mathrm{Z}]$ means the matrix made up of all columns of Ξ and Z including multiplicities, and $\Xi\backslash\mathrm{Z}$ means the matrix obtained from Ξ by omitting all columns in Z up to the multiplicity with which they appear in Z.

In contrast, if H is a *subset* of $\mathbb{R}^s$, then

$$\Xi\backslash H \qquad \text{resp.} \qquad \Xi \cap H$$

is the matrix obtained from Ξ by omitting all columns which do, resp. do not, lie in H.

We say that $\mathrm{Z} \subseteq \Xi$ **spans** in case $\operatorname{ran}\mathrm{Z} = \operatorname{ran}\Xi$, and call Z a **basis** in case it is minimally spanning, i.e., in case no proper submatrix of Z spans. The set of all such bases we denote by

$$\mathcal{B}(\Xi).$$

We use the abbreviation

$$\mathcal{A}(\Xi) := \{\mathrm{Z} \subseteq \Xi : \ \Xi\backslash\mathrm{Z} \text{ does not span}\}$$

for the collection of all $\mathrm{Z} \subseteq \Xi$ which intersect every $B \in \mathcal{B}(\Xi)$, and observe that

$$(16) \qquad \mathrm{Z} \in \mathcal{A}(\Xi \cup \zeta) \quad \Longrightarrow \quad (\mathrm{Z}\backslash\zeta) \in \mathcal{A}(\Xi),$$

since, for any such Z, $\mathrm{Z}\backslash\zeta$ must meet any basis in $\Xi \cup \zeta$ which does not contain ζ, i.e., any basis in Ξ.

The minimal elements of $\mathcal{A}(\Xi)$ are the matrices of the form $\Xi\backslash H$, with H any **hyperplane** (i.e., $(d-1)$-dimensional subspace) spanned by columns of Ξ. We denote by

$$\mathbb{H}(\Xi)$$

the collection of all such hyperplanes and by

$$\mathcal{A}_{\min}(\Xi) := \{\Xi\backslash H : \ H \in \mathbb{H}(\Xi)\}$$

the collection of all minimal elements of $\mathcal{A}(\Xi)$.

In accordance with our convention of treating Ξ as a (multi)set, we use the columns of Ξ also as indices. Thus, if $\mathrm{Z} \subseteq \Xi$, then $f \in T^{\mathrm{Z}}$ indicates that f is a vector indexed by $\zeta \in \mathrm{Z}$ (and with values in the set T). For example, if $f \in \mathbb{R}^{\mathrm{Z}}$, then we feel free to write $\mathrm{Z}f$ for the particular linear combination $\sum_{\zeta\in \mathrm{Z}} \zeta\, f(\zeta)$ of the columns of Z.

Example. As a simple illustration of these conventions, consider

$$\Xi = \begin{bmatrix} 1 & 0 & 1 & 1 & 2 \\ 0 & 1 & 0 & 1 & 2 \end{bmatrix}.$$

Then $s = 2$, $n = 5$, and $d(\Xi) = 2$. Further, for $\mathrm{Z} = \begin{bmatrix} 1 & 1 \\ 0 & 1 \end{bmatrix}$, we have $\mathrm{Z} \subseteq \Xi$, and

$$\Xi \cup \mathrm{Z} = \begin{bmatrix} 1 & 0 & 1 & 1 & 2 & 1 & 1 \\ 0 & 1 & 0 & 1 & 2 & 0 & 1 \end{bmatrix}, \qquad \Xi\backslash \mathrm{Z} = \begin{bmatrix} 0 & 1 & 2 \\ 1 & 0 & 2 \end{bmatrix}.$$

By contrast, with $\mathbf{i}_j$ the jth unit vector and $H := \operatorname{span} \mathbf{i}_1$,

$$\Xi\backslash H = \begin{bmatrix} 0 & 1 & 2 \\ 1 & 1 & 2 \end{bmatrix}, \qquad \Xi \cap H = \begin{bmatrix} 1 & 1 \\ 0 & 0 \end{bmatrix}.$$

The (multi)set $\mathcal{B}(\Xi)$ of all bases in Ξ is given by

$$\begin{bmatrix} 1 & 0 \\ 0 & 1 \end{bmatrix}, \begin{bmatrix} 1 & 1 \\ 0 & 1 \end{bmatrix}, \begin{bmatrix} 1 & 2 \\ 0 & 2 \end{bmatrix}, \begin{bmatrix} 0 & 1 \\ 1 & 0 \end{bmatrix}, \begin{bmatrix} 0 & 1 \\ 1 & 1 \end{bmatrix}, \begin{bmatrix} 0 & 2 \\ 1 & 2 \end{bmatrix}, \begin{bmatrix} 1 & 1 \\ 0 & 1 \end{bmatrix}, \begin{bmatrix} 1 & 2 \\ 0 & 2 \end{bmatrix}.$$

These are eight of the possible ten submatrices of Ξ of order 2. The collection $\mathbb{H}(\Xi)$ consists of just three lines in this case, the line spanned by $\mathbf{i}_1$, by $\mathbf{i}_2$, and by $\mathbf{i}_1 + \mathbf{i}_2$. Correspondingly, the elements of $\mathcal{A}_{\min}(\Xi)$ are the matrices

$$\begin{bmatrix} 0 & 1 & 2 \\ 1 & 1 & 2 \end{bmatrix}, \begin{bmatrix} 1 & 1 & 1 & 2 \\ 0 & 0 & 1 & 2 \end{bmatrix}, \begin{bmatrix} 1 & 0 & 1 \\ 0 & 1 & 0 \end{bmatrix}.$$

Basic properties. The definition (1) readily implies that M_Ξ is nonnegative and that its (closed) support consists of the set

$$\Xi\square = \{\sum_{\xi\in\Xi} \xi t_\xi : 0 \le t_\xi \le 1\},$$

with $\square := [0\,..\,1]^n$. It is less obvious that M_Ξ is a piecewise polynomial function on $\operatorname{ran}\Xi$ (which usually is $\mathbb{R}^s$), of degree $\le k = k(\Xi) = \#\Xi - d(\Xi)$ and in $C^{(m-1)}(\operatorname{ran}\Xi)$, with

$$m := m(\Xi) := \min\{\#\mathrm{Z} : \mathrm{Z} \in \mathcal{A}(\Xi)\} - 1.$$

These facts will be verified using identities for differentiating M_Ξ (cf. (37) below). Note, though, that, even as a function on $\operatorname{ran}\Xi$, $M_\Xi(x)$ is well defined only if M_Ξ is continuous at x.

Fourier transform. By taking $\varphi = \exp(-iy\cdot)$ in (1), we obtain the Fourier transform of M_Ξ as

$$\widehat{M_\Xi}(y) = \prod_{\xi\in\Xi} \frac{1 - \exp(-i\xi y)}{i\xi y}. \tag{17}$$

This shows that the **convolution** of two box splines yields again a box spline,

$$(18) \qquad M_\Xi * M_Z = M_{\Xi \cup Z},$$

and this includes the inductive definition (8) as a special case.

The expression (17) for the Fourier transform of M_Ξ stresses the analogy with univariate cardinal splines. In particular, the univariate B-spline

$$M_n$$

with equally spaced knots $\{0, 1, \ldots, n\}$ is the special case corresponding to

$$\Xi = [1\ 1\ \ldots\ 1] \in \mathbb{R}^{1 \times n}$$

since $\widehat{M}_{[1\ 1\ \ldots\ 1]}(y) = \big((1 - \exp(-iy))/iy\big)^n$.

More generally, if Ξ consists of the unit vectors $\mathbf{i}_\nu \in \mathbb{R}^s$ with multiplicities $r(\nu)$, then

$$M_\Xi(x) = \prod_{\nu=1}^{s} M_{r(\nu)}(x(\nu))$$

is the tensor product B-spline.

Symmetries. Multiplying some of the vectors in Ξ by -1 can be interpreted as a shift of the box spline. This is easily deduced from the formula for the Fourier transform. Let $\varepsilon \in \{-1, 1\}^\Xi$ and set

$$\Xi_\varepsilon := \Xi \operatorname{diag} \varepsilon = [\ldots, \varepsilon(\xi)\xi, \ldots].$$

Since

$$\frac{1 - \exp(-i(-\xi)y)}{i(-\xi)y} = \exp(i\xi y)\frac{1 - \exp(-i\xi y)}{i\xi y},$$

we see from (17) that

$$(19) \qquad \widehat{M}_{\Xi_\varepsilon} = \widehat{M}_\Xi \prod_{\varepsilon(\xi)=-1} \exp(i\xi\cdot).$$

Therefore,

$$(20) \qquad M_{\Xi_\varepsilon} = M_\Xi(\cdot + \sum_{\varepsilon(\xi)=-1} \xi).$$

This leads us to the **centered** box spline

$$(21) \qquad M_\Xi^c := M_\Xi(\cdot + c_\Xi)$$

with **center**

$$c_\Xi := \sum_{\xi\in\Xi} \xi/2.$$

For, (20) implies that M_Ξ^c is invariant under all such changes, i.e.,

$$(22) \qquad M_{\Xi_\varepsilon}^c = M_\Xi^c,$$

while (19) implies that $\widehat{M_\Xi^c}$ is real. In fact,

$$\widehat{M_\Xi^c}(y) = \prod_{\xi\in\Xi} \operatorname{sinc}(\xi y), \qquad \operatorname{sinc}(t) := \sin(t/2)/(t/2).$$

A symmetry of a more obvious sort occurs when $Q\Xi$ is the image of Ξ under some invertible linear map Q on $\mathbb{R}^s$. Since $\widehat{f\circ Q} = \widehat{f}\circ(Q^{-1})^T/|\det Q|$, we obtain from (17) that

$$(23) \qquad M_\Xi = |\det Q| M_{Q\Xi}\circ Q.$$

The same identity holds for the centered box spline. Therefore,

$$(24) \qquad M_\Xi^c = M_{Q\Xi}^c$$

for any linear transformation Q for which $Q\Xi = \Xi_\varepsilon$ for some $\varepsilon \in \{-1,1\}^\Xi$. For an example, see (V.22).

Local structure and truncated power. We will need later the fact that some polynomial pieces which form M_Ξ are of exact degree $\#\Xi - \dim \operatorname{ran}\Xi$. This can be seen most easily by considering the structure of M_Ξ near an extreme point e of its support.

Let e be an **extreme point** of the convex polytope $\operatorname{supp} M_\Xi = \Xi\square$; i.e., e is a proper vertex of the polytope. Then there exists a half space that only meets $\Xi\square$ at e; in other words, there exists y so that $y(x-e) < 0$ for all $x \in (\Xi\square)\backslash e$. Since e is necessarily the image under Ξ of an extreme point of $\square$, e is necessarily of the form $e =: \sum_{\zeta\in Z} \zeta$ for some $Z \subseteq \Xi$, and therefore $x - e \in (\Xi\backslash Z)\square - Z\square$. In particular, the columns of Ξ appear in the last set with sign

$$\sigma(\xi) := \begin{cases} -1, & \xi \in Z; \\ 1, & \xi \in \Xi\backslash Z. \end{cases}$$

Thus, $y\xi < 0$ for all $\xi \in \Xi_e := \Xi_\sigma$, and therefore, from (20),

$$(25) \qquad M_\Xi(\cdot + e) = M_{\Xi_e}.$$

This shows that it is sufficient to consider the special case that $e = 0$ and, for some $y \in \mathbb{R}^s$, $\varepsilon := \max\{y\xi : \xi \in \Xi\} < 0$.

In this case, $y\Xi t = \sum_\xi t_\xi y\xi \le \varepsilon \sum_\xi t_\xi < \varepsilon$ for $t \in \mathbb{R}^n_+ \backslash \square$ (since $\varepsilon < 0$ and, for such t, $\sum_\xi t_\xi \ge \max_\xi t_\xi > 1$), so we find that

$$\langle M_\Xi, \varphi\rangle = \int_\square \varphi(\Xi t)dt = \int_{\mathbb{R}^n_+} \varphi(\Xi t)dt =: \langle T_\Xi, \varphi\rangle$$

for all test functions φ with support in the halfspace $\{x \in \mathbb{R}^s : yx \ge \varepsilon\}$. As 0 is in the interior of this halfspace, it follows that $M_\Xi = T_\Xi$ near 0, with T_Ξ the **truncated power**.

The truncated power T_Ξ is closely related to the box spline M_Ξ. It is defined, for any matrix Ξ for which 0 is an extreme point for $\Xi\square$, as the distribution given by the rule

$$(26) \qquad \langle T_\Xi, \varphi\rangle = \int_{\mathbb{R}^n_+} \varphi(\Xi t)dt, \quad \forall \text{ compactly supported } \quad \varphi \in C(\mathbb{R}^s).$$

Since $\ker \Xi \cap \mathbb{R}^n_+ = 0$ in case 0 is an extreme point of $\Xi\square$, we find that $(y + \ker \Xi) \cap \mathbb{R}^n_+$ is a bounded set for each $y \perp \ker \Xi$. Thus, the same analysis that led to the geometric description (3) of the box spline yields

$$(27) \qquad \langle T_\Xi, \varphi\rangle = \int_{\operatorname{ran} \Xi} \varphi(x) \operatorname{vol}_{n-d}\Big(\Xi^{-1}\{x\} \cap \mathbb{R}^n_+\Big) dx / |\det \Xi|$$

when $\square$ is replaced by $\mathbb{R}^n_+$. This shows that T_Ξ can be identified with the function

$$T_\Xi(x) = \operatorname{vol}_{n-d}\Big(\Xi^{-1}\{x\} \cap \mathbb{R}^n_+\Big)/|\det \Xi|, \qquad x \in \operatorname{ran} \Xi.$$

For the present, we only wish to observe that T_Ξ is homogeneous of degree $\#\Xi - \dim \operatorname{ran} \Xi = n - d$ as a function on $\operatorname{ran} \Xi$. Indeed, $u^n \langle T_\Xi, \varphi\rangle = \langle T_\Xi, \varphi(\cdot/u)\rangle$ for $u > 0$, by (26), therefore, with (27),

$$u^n \int_{\operatorname{ran} \Xi} \varphi(x) T_\Xi(x) dx = \int_{\operatorname{ran} \Xi} \varphi(x/u) T_\Xi(x) dx = u^d \int_{\operatorname{ran} \Xi} \varphi(x) T_\Xi(ux) dx.$$

Hence, we have proved

(28) Proposition. *At every extreme point e of its support, $M_\Xi(\cdot + e)$ is homogeneous of exact degree $k = k(\Xi) = \#\Xi - \dim \operatorname{ran} \Xi$.*

See also (33).

The fact that $M_\Xi = T_\Xi$ near the extreme point 0 of $\Xi\square$ can also be seen from the remarkable formula (recall the abbreviation $\nabla_\Xi := \prod_{\xi\in\Xi}\nabla_\xi$)

$$M_\Xi = \nabla_\Xi T_\Xi, \tag{29}$$

which is an immediate consequence of (26): Indeed, by (26) and for any $\xi \in \Xi$, we have (writing $\mathbb{R}^\Xi$ instead of $\mathbb{R}^n$ to stress the dependence on individual columns of Ξ)

$$\langle T_\Xi(\cdot - \xi), \varphi\rangle = \int_{\mathbb{R}_+^\Xi} \varphi(\Xi t + \xi)dt = \int_{\mathbb{R}_+^\Xi} \varphi(\Xi(t + \mathbf{i}_\xi))dt = \int_{\mathbb{R}_+^\Xi + \mathbf{i}_\xi} \varphi(\Xi t)dt,$$

while $\mathbb{R}_+^\Xi \backslash (\mathbb{R}_+^\Xi + \mathbf{i}_\xi) = \mathbb{R}_+^{\Xi\backslash\xi} \times [0\,..\,1)^{\{\xi\}}$, hence

$$\langle \nabla_\xi T_\Xi, \varphi\rangle = \langle T_\Xi, \varphi\rangle - \langle T_\Xi(\cdot - \xi), \varphi\rangle = \int_{\mathbb{R}_+^{\Xi\backslash\xi}\times[0..1)^{\{\xi\}}} \varphi(\Xi t)dt,$$

and therefore

$$\langle \nabla_\Xi T_\Xi, \varphi\rangle = \int_{[0..1)^\Xi} \varphi(\Xi t)dt = \langle M_\Xi, \varphi\rangle.$$

Differentiation. Denote by $D_\xi := \sum_{\nu=1}^s \xi(\nu)D_\nu$ the derivative in the direction ξ and by $\Delta_\xi\varphi := \varphi(\cdot+\xi)-\varphi$, $\nabla_\xi\varphi := \varphi-\varphi(\cdot-\xi)$ the corresponding forward and backward difference operators.

The formulas

$$\widehat{D_\xi M}(y) = iy\xi\widehat{M}(y), \qquad (M(\cdot - \xi))\widehat{\ }(y) = \exp(-i\xi y)\widehat{M}(y)$$

allow us to conclude from (17) that $D_\xi M_\Xi = \nabla_\xi M_{\Xi\backslash\xi}$ for $\xi \in \Xi$. More generally,

$$D_Z M_\Xi = \nabla_Z M_{\Xi\backslash Z} \quad \text{for } Z \subseteq \Xi \tag{30}$$

(recall $D_Z := \prod_{\zeta\in Z} D_\zeta$ and $\nabla_Z := \prod_{\zeta\in Z}\nabla_\zeta$). In particular,

$$M_\Xi * (D_\Xi f) = (D_\Xi M_\Xi) * f = (\nabla_\Xi M_{[]}) * f = \nabla_\Xi f,$$

with $*$ indicating convolution. Consequently, for smooth f,

$$(\nabla_\Xi f)(t) = \int M_\Xi(t-u) D_\Xi f(u)du. \tag{31}$$

This implies in particular that

$$\ker D_\Xi \subseteq \ker \nabla_\Xi = \ker \Delta_\Xi. \tag{32}$$

As an application of (30), recall from (25) that, for any extreme point e of the support of M_Ξ, $M_\Xi(\cdot+e) = M_{\Xi_e}$. Then, with (30),

$$D_{\Xi_e\backslash Z} M_\Xi(\cdot+e) = D_{\Xi_e\backslash Z} M_{\Xi_e} = \nabla_{\Xi_e\backslash Z} M_Z,$$

while $\nabla_{\Xi_e\backslash Z} M_Z = M_Z$ near the origin since $\operatorname{supp} M_Z(\cdot - \sum_{\zeta\in Z'}\zeta) = Z\square + \sum_{\zeta\in Z'}\zeta$ does not meet the origin for any (nonempty) $Z' \subseteq \Xi_e$. Consequently,

(33) $$D_{\Xi_e\backslash Z} M_\Xi(\cdot+e) = M_Z \qquad \text{near } 0.$$

With the choice $Z \in \mathcal{B}(\Xi)$, this shows that M_Ξ contains polynomial pieces of degree $\geq k(\Xi) = \#\Xi - d(\Xi)$, a weaker result than (28)Proposition.

Arbitrary differentiation of a box spline is most easily handled in terms of the derivatives D_Z with $Z \subseteq \Xi$. The following lemma provides an expression for the general rth order derivative in terms of certain of the derivatives D_Z.

(34)Lemma. *For any $w \in \operatorname{ran}\Xi$, there exist constants a_Z, $Z \subseteq \Xi$, so that*

(35) $$D_w^r := (D_w)^r = \sum_{\#Z=r,\ Z\notin\mathcal{A}(\Xi)} a_Z D_Z + \sum_{\#Z\leq r,\ Z\in\mathcal{A}(\Xi)} a_Z D_w^{r-\#Z} D_Z.$$

Proof. Since any $w \in \operatorname{ran}\Xi$ has the form $w = \sum_{\xi\in\Xi} a_\xi \xi$, we can always write D_w^r in the form $\sum_Z a_Z D_Z$, with the sum over all matrices Z with $\#Z = r$ and with columns taken from Ξ. But our goal here is more subtle: We want to write D_w^r in terms of D_Z with $Z \subseteq \Xi$.

For this, we use induction on r and, for the induction step, apply D_w to both sides of (35). The terms in the second sum present no problems. As to the first sum, $w = \sum_{\xi\in\Xi\backslash Z} a_\xi \xi$ for suitable a_ξ, since, by definition of $\mathcal{A}(\Xi)$, $\Xi\backslash Z$ contains a basis. Therefore,

$$D_w D_Z = \sum_{\xi\in\Xi\backslash Z} a_\xi D_\xi D_Z = \Big(\sum_{Z\cup\xi\notin\mathcal{A}(\Xi)} + \sum_{Z\cup\xi\in\mathcal{A}(\Xi)}\Big) a_\xi D_{\xi\cup Z}$$

which yields (35) with r replaced by $r+1$. ☐

It is not possible to replace $\mathcal{A}(\Xi)$ in (35) by $\mathcal{A}_{\min}(\Xi)$, not even just in the second sum, since there may be $Z \notin \mathcal{A}(\Xi)$ and $\xi \in \Xi\backslash Z$ so that $(Z\cup\xi) \in \mathcal{A}(\Xi)\backslash\mathcal{A}_{\min}(\Xi)$. A simple example is provided by the submatrix $Z := [\mathbf{i}_1, \mathbf{i}_2] \subset \Xi := [\mathbf{i}_1, \mathbf{i}_1, \mathbf{i}_2, \mathbf{i}_2]$ for which the two possible choices for $\xi \in \Xi\backslash Z$ both give an element of $\mathcal{A}(\Xi)$, but not one of the two possible elements of $\mathcal{A}_{\min}(\Xi) = \{[\mathbf{i}_1, \mathbf{i}_1], [\mathbf{i}_2, \mathbf{i}_2]\}$.

Proof of basic properties. With the aid of (34)Lemma, we can now derive the basic properties of the box spline stated earlier.

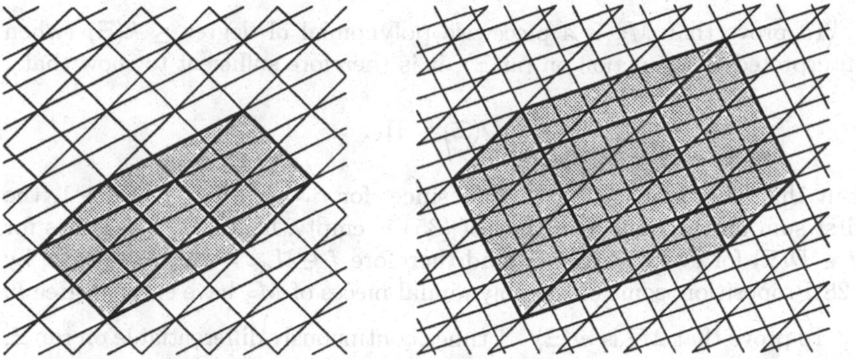

(36)Figure. $\Gamma(\Xi)$ for $\Xi = \begin{bmatrix} 1 & -1 & 2 \\ 1 & 1 & 1 \end{bmatrix}$ and $\Xi = \begin{bmatrix} 3 & -1 & 5 \\ 1 & 4 & 1 \end{bmatrix}$. The mesh $\Gamma_{\text{loc}}(\Xi)$ for M_Ξ is much simpler.

(37)Proposition.

(i) *As a function on* $\operatorname{ran}\Xi$, M_Ξ *is piecewise in*

$$D(\Xi) := \bigcap_{Z\in\mathcal{A}(\Xi)} \ker D_Z = \bigcap_{Z\in\mathcal{A}_{\min}(\Xi)} \ker D_Z$$

on the complement of the **local mesh**

$$\Gamma_{\text{loc}}(\Xi) := \{\Xi t : \ \{\xi \in \Xi : t_\xi \in \{0,1\}\} \in \mathcal{A}(\Xi) \ \}.$$

Further, $D(\Xi) \subseteq \Pi_k(\operatorname{ran}\Xi)$, *with* $k = k(\Xi) = \#\Xi - d(\Xi)$.

(ii) $M_\Xi \in C^{(m-1)}(\operatorname{ran}\Xi)$, *with* $m = m(\Xi) = \min\{\#Z : Z \in \mathcal{A}(\Xi)\} - 1$.

Proof. For any $Z \in \mathcal{A}(\Xi)$ (as for any $Z \subseteq \Xi$), $D_Z M_\Xi = \nabla_Z M_{\Xi\backslash Z}$ (by (30)) is a linear combination of shifts of the box spline $M_{\Xi\backslash Z}$. Since

$$\nabla_Z\varphi = \sum_{Y\subseteq Z}(-1)^{\#Y}\varphi(\cdot - \sum_{\zeta\in Y}\zeta),$$

the support of $\nabla_Z M_{\Xi\backslash Z}$ is contained in the set

$$\bigcup_{Y\subseteq Z}\big(\sum_{\zeta\in Y}\zeta + (\Xi\backslash Z)\square\big),$$

and this set is in $\Gamma_{\text{loc}}(\Xi)$ since $Z \in \mathcal{A}(\Xi)$. Therefore $D_Z M_\Xi$ vanishes on the complement of Γ_{loc}. Since the set Γ_{loc} is of lower dimension in $\operatorname{ran}\Xi$, it follows that M_Ξ agrees with some element of $D(\Xi)$ on any connected subset of $(\operatorname{ran}\Xi)\backslash\Gamma_{\mathrm{loc}}$. Since $Z \subseteq Z'$ implies that $\ker D_Z \subseteq \ker D_{Z'}$, $D(\Xi)$ is already the intersection of the kernels of D_Z for all *minimal* $Z \in \mathcal{A}(\Xi)$.

To prove that M_Ξ is a piecewise polynomial of degree $\le k(\Xi)$ (when interpreted as a function on $\operatorname{ran}\Xi$), it is therefore sufficient to show that

$$D(\Xi) \subseteq \Pi_k.$$

But this follows at once from (35) since, for $r = k+1 = n-d+1$, the first sum on the right hand side of (35) is empty, hence $D_w^r f$ vanishes for $f \in D(\Xi)$ for every $w \in \operatorname{ran}\Xi$, and therefore $f \in \Pi_{<r} = \Pi_k$. Note that, by (28)Proposition, some of the polynomial pieces of M_Ξ have exact degree k.

To prove that M_Ξ is $m(\Xi)-1$ times continuously differentiable on $\operatorname{ran}\Xi$, we note that, for $w \in \operatorname{ran}\Xi$ and $r = m < \min\{\#Z : Z \in \mathcal{A}(\Xi)\}$, the second sum on the right hand side of (35) is empty. Therefore, with (30), $D_w^r M_\Xi$ is a linear combination of shifts of the box splines $M_{\Xi\backslash Z}$ with $Z \notin \mathcal{A}(\Xi)$. Since, for such Z, $\Xi\backslash Z$ spans, these box splines are bounded functions on $\operatorname{ran}\Xi$, and it follows that all derivatives of M_Ξ of order m and in directions in $\operatorname{ran}\Xi$ are bounded. ☐

The first assertion of the proposition implies that the only discontinuities of M_Ξ occur at points in Γ_{loc}. At these points, M_Ξ has at least $m-1$ continuous derivatives. But, depending on Ξ there may be points in Γ_{loc} at which M_Ξ is smoother than that. For example, for the bivariate box spline M_Ξ with $\Xi = [\mathbf{i}_1, \mathbf{i}_1, \mathbf{i}_2]$, $m(\Xi) = 0$ and hence M_Ξ has jumps, but it is continuous across mesh lines parallel to $\mathbf{i}_2$. (38)Figure shows another such example.

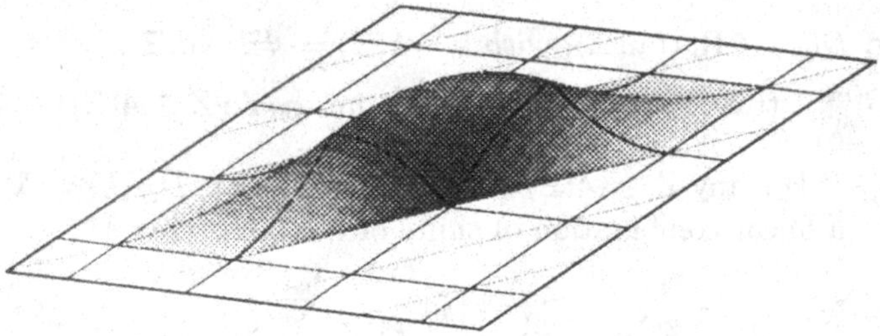

(38)Figure. The box spline $M_{\left[\begin{smallmatrix}1&0&1&1\\0&1&1&1\end{smallmatrix}\right]}$ is only $C^{(0)}$ across diagonals, but is otherwise $C^{(1)}$.

As indicated in (36)Figure, the integer translates of $\Gamma_{\mathrm{loc}}(\Xi)$ make up the **finer mesh**

$$\Gamma(\Xi) := \bigcup_{H \in \mathbb{H}(\Xi)} H + \mathbb{Z}^s.$$

This means that any linear combination of integer translates of the box spline coincides, on each connected component of this finer mesh, with some polynomial in $D(\Xi)$. Note that $\Gamma(\Xi)$ is dense in case the columns of Ξ fail to be commensurate.

(39)Example. For $\Gamma(\Xi)$ to be simple, Ξ has to be chosen with considerable care. For example, if $\Gamma(\Xi)$ is to be a square mesh in the plane, then, up to an affine change of variables, the columns of Ξ must be taken from $\{\pm\mathbf{i}_1, \pm\mathbf{i}_2\}$. If also the SW-to-NE diagonal is allowed as a mesh-line, then some columns can be of the form $\pm(\mathbf{i}_1 + \mathbf{i}_2)$. Finally, if both diagonals are permitted, then, in addition, some columns can be of the form $\pm(\mathbf{i}_1 - \mathbf{i}_2)$. The resulting three meshes are shown in (40)Figure. □

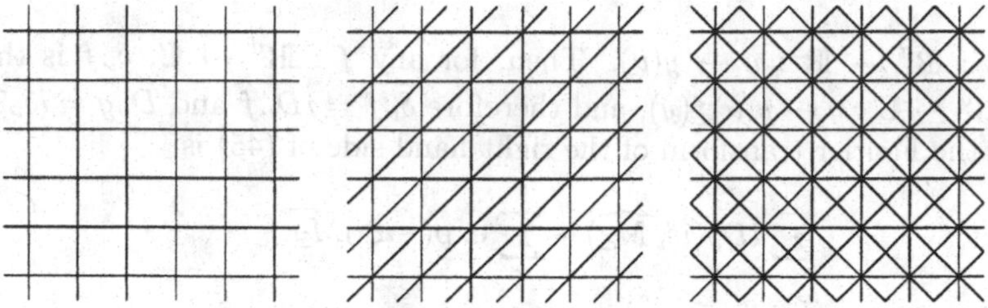

(40)Figure. The two-, three-, and four-direction meshes.

The following monotonicity properties of $\Xi \mapsto D(\Xi)$ will be of use later.

(41)Proposition. *If* $\Xi \subseteq \Xi'$ *and* $\operatorname{ran}\Xi = \operatorname{ran}\Xi'$, *then*

$$D_{\Xi'\backslash\Xi} D(\Xi') \subseteq D(\Xi) \subseteq D(\Xi'). \tag{42}$$

Proof. If $Z \in \mathcal{A}(\Xi)$, then $Z' := (\Xi'\backslash\Xi) \cup Z$ is certain to meet every basis in Ξ', hence is in $\mathcal{A}(\Xi')$. Therefore, if $f \in D_{\Xi'\backslash\Xi}D(\Xi')$, hence $f = D_{\Xi'\backslash\Xi}g$ for some $g \in D(\Xi')$, then $D_Z f = D_{Z'} g = 0$. This proves the first inclusion in (42).

For the second inclusion, note that any $Z' \in \mathcal{A}(\Xi')$ is certain to meet every $B \in \mathcal{B}(\Xi)$, hence is of the form $Z' = Z'' \cup Z$, with $Z'' \subseteq \Xi'\backslash\Xi$ and $Z \in \mathcal{A}(\Xi)$. Therefore, if $f \in D(\Xi)$, then $D_{Z'} f = D_{Z''} D_Z f = 0$. □

We now come to the last basic box spline property to be discussed here.

(43)Recurrence Relation. *If the box splines* $M_{\Xi\backslash\xi}$, $\xi \in \Xi$, *are continuous at* $x = \Xi t$ *(as functions on* $\operatorname{ran}\Xi = \mathbb{R}^d$*), then*

$$(n-d)M_\Xi(x) = \sum_{\xi\in\Xi} t_\xi M_{\Xi\backslash\xi}(x) + (1-t_\xi)M_{\Xi\backslash\xi}(x-\xi). \tag{44}$$

While t can always be chosen from $\square$ when $x \in \operatorname{supp} M_\Xi$, t need not be so chosen.

Proof. By a change of variables, using (23), we may assume that $\operatorname{ran} \Xi = \mathbb{R}^d$. Then, by (30),

$$DM_\Xi(x) := (D_x M_\Xi)(x) = \sum_{\xi \in \Xi} t_\xi \big(M_{\Xi\backslash\xi}(x) - M_{\Xi\backslash\xi}(x - \xi) \big),$$

and it remains to show that

$$(n - d) M_\Xi = DM_\Xi + \sum_\xi M_{\Xi\backslash\xi}(\cdot - \xi). \tag{45}$$

Let $\delta_\nu : \mathbb{R}^d \to \mathbb{R} : y \mapsto y(\nu)$. Then, for any $f : \mathbb{R}^d \to \mathbf{C}$, $\delta_\nu f$ is the map $\mathbb{R}^d \to \mathbf{C} : y \mapsto y(\nu) f(y)$, and therefore $\widehat{\delta_\nu f} = i D_\nu \widehat{f}$ and $\widehat{D_\nu g} = i \delta_\nu \widehat{g}$. Hence the Fourier transform of the right hand side of (45) is

$$\sum_\nu i D_\nu (i \delta_\nu \widehat{M_\Xi}) + \sum_\xi \exp(-i\xi \cdot) \widehat{M_{\Xi\backslash\xi}} \,.$$

Since $D_\nu (\delta_\nu f) = f + \delta_\nu D_\nu f$, the first sum equals

$$(-d) \widehat{M_\Xi} - \sum_\nu \delta_\nu D_\nu \widehat{M_\Xi}.$$

This finishes the proof since, by (17),

$$D_\nu \widehat{M_\Xi}(y) = \sum_\xi \frac{-i\xi(\nu)}{i\xi y} \widehat{M_\Xi}(y) + \frac{i\xi(\nu)}{i\xi y} \exp(-i\xi y) \widehat{M_{\Xi\backslash\xi}}(y),$$

hence

$$\begin{aligned} \sum_\nu \delta_\nu(y) D_\nu \widehat{M_\Xi}(y) &= \sum_\xi \frac{-i\xi y}{i\xi y} \widehat{M_\Xi}(y) + \frac{i\xi y}{i\xi y} \exp(-i\xi y) \widehat{M_{\Xi\backslash\xi}}(y) \\ &= \; -n \, \widehat{M_\Xi}(y) \; + \sum_\xi \exp(-i\xi y) \widehat{M_{\Xi\backslash\xi}}(y). \end{aligned}$$

$\square$

Since M_Ξ is piecewise polynomial, the recurrence relation is applicable to *arbitrary* x, with the proviso that, in case of discontinuities, the value at x of the various $M_{\Xi\backslash\xi}$ be chosen consistently. This can be accomplished, once x is recognized to be a point of discontinuity, by perturbing x randomly by a 'small' amount.

(46)Example: four ways to construct a box spline. Consider the **ZP** element, i.e., the bivariate quadratic box spline corresponding to the matrix $\Xi = \begin{bmatrix} 1 & 0 & 1 & -1 \\ 0 & 1 & 1 & 1 \end{bmatrix}$. Since every pair of columns of Ξ spans, $m(\Xi) = 2$, and hence M_Ξ is continuously differentiable with possible discontinuities in the second derivatives across the mesh-lines shown in (47)Figure. In fact, the dotted mesh lines are not active because the ZP element is in $C^{(1)}$ and is even across each such line, hence $C^{(2)}$ across such a line, which means that the quadratic polynomial pieces joining across such a line must come from the same polynomial.

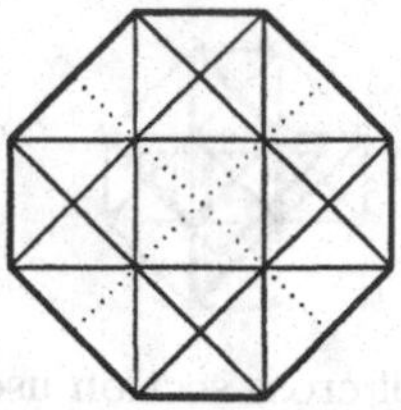

(47)Figure. Grid lines for the ZP element.

To illustrate the basic definitions and identities of this section, we discuss four different possibilities for computing this particular box spline M_Ξ:

(i) The set $\Xi^{-1}\{x\} \cap \square$ in definition (3) is the intersection of $[0\,..\,1)^4$ with the plane

$$V := t_0 + \ker \Xi = \{t_0 + \mathrm{K}y : y \in \mathbb{R}^2\}$$

where t_0 is any vector with $x = \Xi t_0$, and K is a matrix whose columns form a basis for $\ker \Xi$, e.g., $\mathrm{K} = \begin{bmatrix} -1 & -1 & 1 & 0 \\ 1 & -1 & 0 & 1 \end{bmatrix}^T$. It is instructive (and possible in this simple case) to visualize $\Xi^{-1}\{x\} \cap \square$ for specific choices of x, as is done in (48)Figure. As x changes, so does the plane V, but it always stays parallel to $\ker \Xi$. This makes it reasonable to plot the outlines of the orthogonal projection onto the plane $\ker \Xi$ of each such set $\Xi^{-1}\{x\} \cap \square$, as is done in (48)Figure; see the shaded areas there. The very same ten values of x are also used in (49)Figure, hence their position relative to the support of the ZP element can be read off from that Figure. (48)Figure also shows the orthogonal projection onto $\ker \Xi$ of the edges and vertices of the unit cube $[0\,..\,1]^4$, with the thickness of a line reflecting the average distance of the corresponding edge from the plane $\ker \Xi$. (There is no pretense here that one could actually visualize the cube $[0\,..\,1]^4$ being intersected by that plane V; still, it is amusing to try.)

The next in the sequence of pictures in the top row of (48)Figure would show the biggest possible shaded area (corresponding to the maximum value of the ZP element), viz., the square with vertices at the midpoints of the horizontal and vertical edges of the projection of the four-cube. In particular, the shaded area will never fill out the entire projection of the four-cube.

For computing $M_\Xi(x)$, the shaded area is measured with respect to orthonormal units (a segment of length two units is shown in (48)Figure)

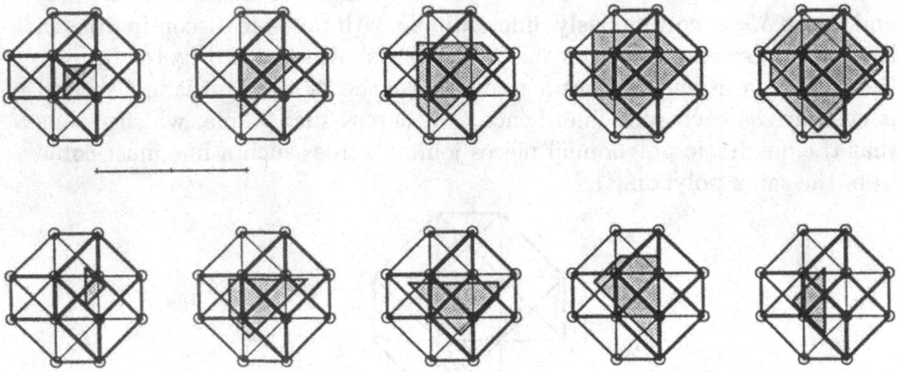

(48)Figure. Various areas of cross section used in the calculation of the ZP element.

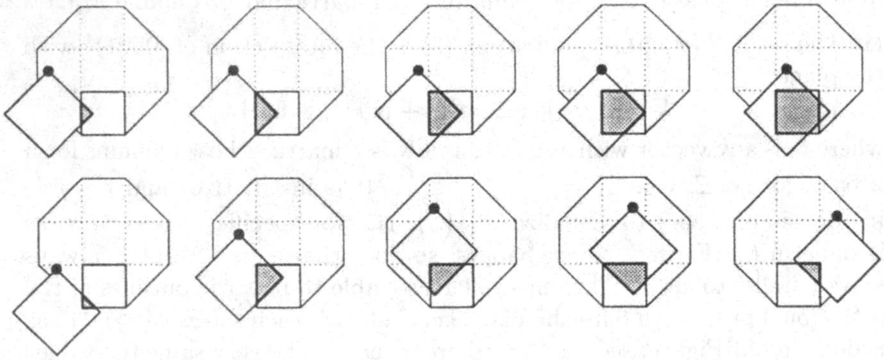

(49)Figure. The ZP element obtained by convolving one square with another. This figure parallels (48)Figure.

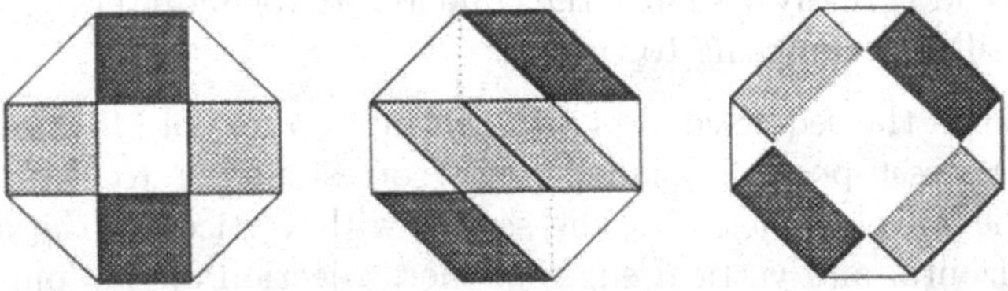

(50)Figure. Derivatives $D_{\{1,2\}}$, $D_{\{2,3\}}$ and $D_{\{3,4\}}$ of the ZP element.

and then divided by

$$|\det \Xi| = \sqrt{\det \Xi\Xi^T} = 3.$$

E.g., in those units, the triangle in the top row, second from left, has side length .85 (to graphic accuracy), hence area .36125, giving the approximate value .36125/3=.12042 for $M(-.5, 1.5)$. In fact, $M(-.5, 1.5) = (1/4)/2 =$ .125, as one can read off from the corresponding second figure in the top row of (49)Figure. Incidentally, comparison of the projection of the four-cube with the 'ruler' in (48)Figure shows that the projection of every edge of the four-cube is foreshortened, consistent with the fact that none of these edges is parallel to $\ker \Xi$. **(ii)** Using (8), M_Ξ can be obtained by averaging, starting, e.g., with the characteristic function χ of the unit square (as already pointed out in (11)Example). Using the notation $\{i, j, \ldots\}$ to refer to the submatrix of Ξ consisting of columns $i, j, \ldots$, we have

$$M_{\{1,2,3\}}(x) = M_{\{3\}} * M_{\{1,2\}}(x) = \int_0^1 \chi(x_1 - t, x_2 - t)\, dt$$

by averaging in the direction $\zeta = (1, 1)$; subsequent averaging, in the direction $\xi = (-1, 1)$, yields

$$M_\Xi(x) = M_{\{3,4\}} * M_{\{1,2\}}(x) = \int_0^1 \int_0^1 \chi(x_1 - t + t', x_2 - t - t')\, dtdt'.$$

Thus, $M_\Xi(x)$ is the area of the intersection of two squares divided by the product of the areas of the two squares, i.e., divided by 2. This is illustrated in (49)Figure. The five x's used in the top row of that Figure are $(.25j, 1.5)$, $j = -3, \ldots, 1$. The five x's used in the bottom row describe a more complicated path through the support of the ZP element. For example, the second figure in the top row shows the situation for $x = (-.5, 1.5)$. It appears that the larger square covers exactly one quarter of the smaller square, hence $M_\Xi(-.5, 1.5) = (1/4)/2 = 1/8$.

(iii) Since M_Ξ has compact support and has continuous first and piecewise continuous second derivatives, we can write $M_\Xi(x)$ as a weighted sum of the second directional derivative $D_w^2 M_\Xi$ over the interval $x + [-\infty \,..\, 0]w$, for arbitrary $w \in \mathbb{R}^s \backslash 0$, i.e.,

$$M_\Xi(x) = M_\Xi(x + tw)_{|t=0} = \int_{-\infty}^0 (-t)\,(D_w^2 M_\Xi)(x + tw)\, dt.$$

This makes use of the fact that $D_w^2 M_\Xi(x + tw)$ is the second derivative at t of the *univariate* function $g : t \mapsto M_\Xi(x + tw)$, hence $M_\Xi(x + sw) = g(a) + Dg(a)(s - a) + \int_a^s (s - t)\, D^2 g(t)\, dt = \int_a^s (s - t)\, D^2 g(t)\, dt$ for any s, and for any $x + aw \notin \Xi\square$. Any such second directional derivative $D_w^2 M_\Xi$ of M_Ξ can be computed with the aid of (30). For example,

$$\begin{aligned} D_{\{1,2\}} M_\Xi &:= D_1 D_2 M_\Xi \\ &= M_{\{3,4\}} - M_{\{3,4\}}(\cdot - (1,0)) - M_{\{3,4\}}(\cdot - (0,1)) + M_{\{3,4\}}(\cdot - (1,1)), \end{aligned}$$

i.e., $D_{\{1,2\}}M_\Xi$ is the sum of shifts of characteristic functions of squares. Since these shifts overlap, $D_{\{1,2\}}M_\Xi$ is actually the disjoint sum of characteristic functions of rectangles; see (50)Figure. Similarly one computes $D_{\{2,3\}}M_\Xi$ and $D_{\{3,4\}}M_\Xi$. The support of these derivatives is shown in (50)Figure with dark (light) shading indicating where $D_{\{i,j\}}M_\Xi$ is positive (negative). Since all derivatives of order 2 can be expressed as linear combinations of $D_{\{1,2\}}$, $D_{\{2,3\}}$ and $D_{\{3,4\}}$, it follows that $D_{\{i,j\}}M_\Xi$ can have discontinuities only across the boundaries of the shaded regions. This shows again that the dotted lines in (47)Figure are not mesh lines for this M_Ξ.

(iv) (51)Figure illustrates the computation of M_Ξ via the recurrence relation (44). In general, there is no unique choice of the weights t_ξ. For efficiency, one represents x in terms of a basis, so that only two of the t's are nonzero at each stage of the recurrence. This might require some t_ξ to lie outside $[0\,..\,1]$ (as is permitted by (43)). Further, one notes that the only branches of the tree which contribute to the computation are those for which x lies in the support of the corresponding *shifted* box splines. In the figure, x is chosen in the darkest triangle and, with appropriate choice of the weights, the recurrence only involves $M_{\{1,2,4\}}$, and shifts of $M_{\{1,2,3\}}$, $M_{\{1,3,4\}}$ and $M_{\{2,3,4\}}$. More precisely,

$$\begin{aligned}(4-2)M_{\{1,2,3,4\}}(x) &= M_{\{2,3,4\}}(x-(1,0)) + M_{\{1,3,4\}}(x-(0,1))\\ &\quad + t_{\{3\}}M_{\{1,2,4\}}(x) + (1-t_{\{4\}})M_{\{1,2,3\}}(x-(-1,1)).\end{aligned}$$

Carrying the recursion one step further, one finds

$$\begin{aligned}M_{\{2,3,4\}}(x-(1,0)) &= t'_{\{3\}}M_{\{2,4\}}(x-(1,0)) + M_{\{2,3\}}(x-(1,0)-(-1,1))\\ &= t'_{\{3\}} + 1\\ M_{\{1,3,4\}}(x-(0,1)) &= t''_{\{1\}}M_{\{3,4\}}(x-(0,1)) + t''_{\{3\}}M_{\{1,4\}}(x-(0,1))\\ &= t''_{\{1\}}/2 + t''_{\{3\}}\\ M_{\{1,2,4\}}(x) &= (1-t'''_{\{2\}})M_{\{1,4\}}(x-(0,1)) + M_{\{2,4\}}(x-(1,0))\\ &= 1 - t'''_{\{2\}} + 1\\ M_{\{1,2,3\}}(x-(-1,1)) &= (1-t''''_{\{1\}})M_{\{2,3\}}(x-(-1,1)-(1,0))\\ &= 1 - t''''_{\{1\}}.\end{aligned}$$

The weights in the above formulas are determined from the equations

$$\begin{aligned}x &= t_{\{3\}}\left[\begin{smallmatrix}1\\1\end{smallmatrix}\right] + t_{\{4\}}\left[\begin{smallmatrix}-1\\1\end{smallmatrix}\right]\\ &= t'_{\{2\}}\left[\begin{smallmatrix}0\\1\end{smallmatrix}\right] + t'_{\{3\}}\left[\begin{smallmatrix}1\\1\end{smallmatrix}\right] + \left[\begin{smallmatrix}1\\0\end{smallmatrix}\right] = t''_{\{1\}}\left[\begin{smallmatrix}1\\0\end{smallmatrix}\right] + t''_{\{3\}}\left[\begin{smallmatrix}1\\1\end{smallmatrix}\right] + \left[\begin{smallmatrix}0\\1\end{smallmatrix}\right]\\ &= \quad t'''_{\{2\}}\left[\begin{smallmatrix}0\\1\end{smallmatrix}\right] + t'''_{\{4\}}\left[\begin{smallmatrix}-1\\1\end{smallmatrix}\right] \quad = t''''_{\{1\}}\left[\begin{smallmatrix}1\\0\end{smallmatrix}\right] + t''''_{\{3\}}\left[\begin{smallmatrix}1\\1\end{smallmatrix}\right] + \left[\begin{smallmatrix}-1\\1\end{smallmatrix}\right].\end{aligned}$$

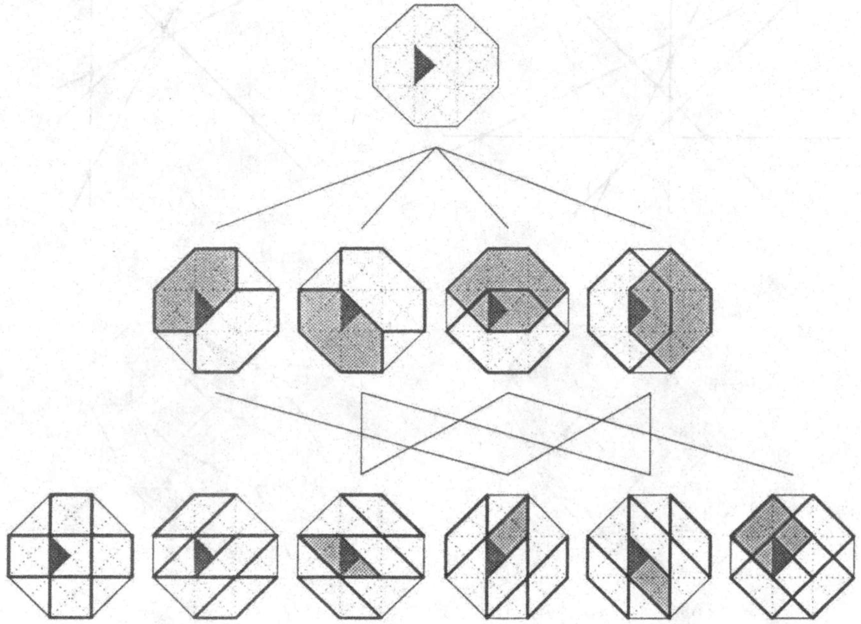

(51)Figure. The ZP element obtained by recurrence.

For example, when $x = (1/4, 3/2)$, we have $t_{\{3\}} = 7/8$, $t_{\{4\}} = 5/8$, $t'_{\{2\}} = 9/4$, $t'_{\{3\}} = -3/4$, $t''_{\{1\}} = -1/4$, $t''_{\{3\}} = 1/2$, $t'''_{\{2\}} = 7/4$, $t'''_{\{4\}} = -1/4$, $t''''_{\{1\}} = 3/4$, $t''''_{\{3\}} = 1/2$ and the recurrence relation yields

$$\begin{aligned} 2M_{\{1,2,3,4\}}(x) &= 1+t'_{\{3\}}+t''_{\{1\}}/2+t''_{\{3\}}+t_{\{3\}}(2-t'''_{\{2\}})+(1-t_{\{4\}})(1-t''''_{\{1\}}) \\ &= 15/16. \end{aligned}$$

A check of sorts is available for this number, since it corresponds to the last figure in the top rows of (48)Figure and (49)Figure which illustrate use of methods (i) and (ii) for evaluating the ZP element. E.g., in (49)Figure, the larger square appears to cut off from the smaller (unit) square two triangles of side length 1/4 each, leaving an area of size $1-(1/4)^2 = 15/16$. A similar check could be carried out on the rightmost figure in the top row of (48)Figure, using the ruler provided with the figure.

The support of the box spline. The inductive definition facilitates the proof of the fact that the support of M_Ξ is the essentially disjoint union of parallelepipeds spanned by bases in Ξ. This fact in itself is perhaps not

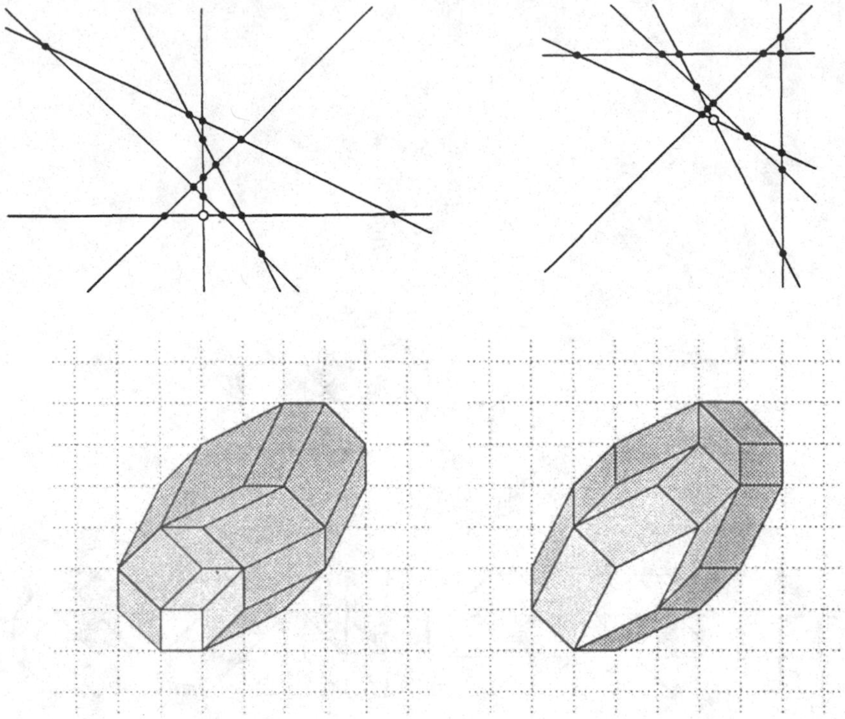

(52) Figure. Two inductive partitions of the support of M_Ξ, for $\Xi = \begin{bmatrix} 1 & 0 & 1 & -1 & 2 & 1 \\ 0 & 1 & 1 & 1 & 1 & 2 \end{bmatrix}$. Note that the correspondence, in which each parallelogram is associated with the intersection of the two lines above which are perpendicular to its two sides, is 1-1. This shows the partition to be the 'zonotope' for the corresponding line arrangement, as explained later.

too surprising since the recurrence relation allows us to write $M_\Xi(x)$ as a sum

$$\sum_{\substack{\mathrm{Z}\subseteq\Xi \\ \#\mathrm{Z}=d}} \sum_{a\in\{0,1\}^{\Xi\backslash \mathrm{Z}}} t_a(x) M_{\mathrm{Z}}(x-(\Xi\backslash \mathrm{Z})a)$$

of lower order box splines each of which, if it is a function on $\operatorname{ran}\Xi$, is the characteristic function of some parallelepiped spanned by some basis in Ξ. What is remarkable is that there is such a decomposition of the support of M_Ξ in which each basis in Ξ occurs in this fashion *exactly once*, counting multiplicities. Here is the precise statement.

(53) Theorem. *Let* $\operatorname{ran}\Xi = \mathbb{R}^s$. *There exist points* $\alpha_{\mathrm{Z}} \in \Xi\{0,1\}^n$, $\mathrm{Z} \in \mathcal{B}(\Xi)$, *so that* $\Xi\square$ *is the essentially disjoint union of the sets*

$$\mathrm{Z}\square + \alpha_{\mathrm{Z}}, \quad \mathrm{Z} \in \mathcal{B}(\Xi).$$

In particular,

$$\mathrm{vol}_s\, \Xi\square = \sum_{Z\in\mathcal{B}(\Xi)} |\det Z|.$$

Proof. The theorem will be proved by induction on $\#\Xi - s$. If $\#\Xi - s = 0$, then Ξ is the only element of $\mathcal{B}(\Xi)$ and the theorem is clear. Assume now that the theorem holds for $\#\Xi - s$ and let $\zeta \in \mathbb{R}^s$ be given. The idea of the proof is as follows: We observe that

$$(\Xi\cup\zeta)\square\backslash\Xi\square = G + (0\,..\,\zeta],$$

with

$$G := \{z \in \Xi\square : \Xi\square\cap[z\,..\,z+\zeta] = z\}.$$

The set G is necessarily part of the boundary of $\Xi\square$. Therefore, almost all its points are associated with exactly one of the hyperplanes in $\mathbb{H}(\Xi)$. Further, these are necessarily those hyperplanes $H \in \mathbb{H}(\Xi)$ for which $\zeta \notin H$. (In fact, each such hyperplane is associated with some boundary points, as is shown later.) Since we already have the desired partition of $\Xi\square$ in terms of $Z \in \mathcal{B}(\Xi)$, and the elements of $\mathcal{B}(\Xi\cup\zeta)\backslash\mathcal{B}(\Xi)$ are precisely $W\cup\zeta$ where W is a basis for some H with $\zeta \notin H$, these remaining elements of $\mathcal{B}(\Xi\cup\zeta)$ will be used to partition $G + (0\,..\,\zeta]$.

Throughout, H or H' denote hyperplanes in $\mathbb{H}(\Xi)$. We begin by deriving a necessary condition for $z \in \Xi\square$ to be a boundary point of $\Xi\square$. If $z \in \Xi\square$, then $z = \sum_{\xi\in\Xi} a_z(\xi)\xi$ with $a_z \in [0\,..\,1]^n$. (For given z, there may be several choices for a_z here, but that doesn't matter.) If $\{\xi \in \Xi : 0 < a_z(\xi) < 1\}$ spans, then a whole neighborhood of z belongs to $\Xi\square$. Hence any boundary point of $\Xi\square$ must lie in a set of the form

$$H + \sum_{\xi\in\Xi\backslash H} a(\xi)\xi \tag{54}$$

for some $H \in \mathbb{H}(\Xi)$ and with $a(\xi) \in \{0, 1\}$ for all $\xi \in \Xi\backslash H$. Note that the collection of boundary points z which belong to two such sets, associated with distinct hyperplanes, H and H' say, is of $(s-1)$-dimensional measure 0 since then $\{\xi : 0 < a_z(\xi) < 1\} \subset H \cap H'$, a set of dimension $< s - 1$. It is therefore sufficient to restrict attention to z in (54) which belong to only one such set, i.e., for which H is spanned by $\{\xi \in \Xi : 0 < a_z(\xi) < 1\}$.

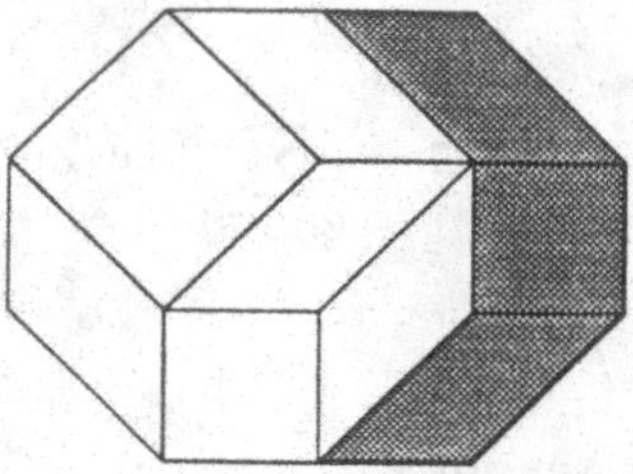

(55)Figure. The support of $M_{\Xi\cup\zeta}$, for $\Xi = [\mathbf{i}_1, \mathbf{i}_2, \mathbf{i}_1 + \mathbf{i}_2, -\mathbf{i}_1 + \mathbf{i}_2]$ and $\zeta = \mathbf{i}_1$.

Next we prove a necessary and sufficient condition for such a point z from (54) to belong to G, i.e., to be a boundary point of $\Xi\square$ at which the direction ζ points outward. We claim that such z belongs to G if and only if $a_z = g_H$ on $\Xi\backslash H$, with

$$(56) \qquad g_H(\xi) := \begin{cases} 1, & \text{if } H^\perp\xi/H^\perp\zeta > 0; \\ 0, & \text{if } H^\perp\xi/H^\perp\zeta < 0, \end{cases}$$

and $H^\perp$ a (nontrivial) normal to H. Indeed, if for a z in (54), $z + b\zeta$ lies in $\Xi\square$, then $z + b\zeta = \sum_{\xi\in\Xi} c(\xi)\xi$ for some $c \in [0\,..\,1]^\Xi$. Hence

$$\sum_{\xi\in\Xi\backslash H} a_z(\xi)H^\perp\xi + bH^\perp\zeta \;=\; H^\perp\,(z + b\zeta) \;=\; \sum_{\xi\in\Xi\backslash H} c(\xi)H^\perp\xi$$

or

$$b = \sum_{\xi\in\Xi\backslash H} \Big(c(\xi) - a_z(\xi)\Big)\frac{H^\perp\xi}{H^\perp\zeta}.$$

Thus the condition $a_z = g_H$ on $\Xi\backslash H$ implies that $b \le 0$, i.e., $z \in G$. Conversely, if this condition is violated (i.e., if, for some $\xi_0 \in \Xi\backslash H$, e.g. $a_z(\xi_0) = 0$ yet $H^\perp\xi_0/H^\perp\zeta > 0$), then $\zeta = h + \frac{H^\perp\zeta}{H^\perp\xi_0}\xi_0$ for some $h \in H$. Therefore, $h = \sum_{0<a_z(\xi)<1} a_h(\xi)\xi$ and

$$z + b\zeta = \sum_{0<a_z(\xi)<1} (a_z(\xi) + ba_h(\xi))\xi + \sum_{a_z(\xi)\in\{0,1\},\xi\neq\xi_0} a_z(\xi)\xi$$
$$+ \Big(a_z(\xi_0) + b\frac{H^\perp\zeta}{H^\perp\xi_0}\Big)\xi_0,$$

and this is in $\Xi\square$ for all sufficiently small positive b, i.e., z fails to lie in G.

We conclude that the boundary of $\Xi\square$ at which ζ points outward, i.e., the set G, is the essentially disjoint union of the sets

$$(\Xi\cap H)\square + \sum_{\xi\in\Xi\backslash H} g_H(\xi)\xi$$

with $H^\perp\zeta \neq 0$, i.e., with $\zeta \notin H$. Consequently, $G+(0\,..\,\zeta] = (\Xi\cup\zeta)\square\backslash\Xi\square$ is the essentially disjoint union of the corresponding sets

$$\Bigl((\Xi\cap H)\square+(0\,..\,\zeta]\Bigr)+\sum_{\xi\in\Xi\backslash H} g_H(\xi)\xi$$

with $H^\perp\zeta \neq 0$, i.e., with $\zeta \notin H$, and induction provides the partition of each of these sets into sets of the form $\mathrm{Z}\square+\alpha_{\mathrm{Z}}$ with each $\mathrm{Z}\in\mathcal{B}((\Xi\cap H)\cup\zeta)$ occurring exactly once. Since each $\mathrm{Z}\in\mathcal{B}(\Xi\cup\zeta)\backslash\mathcal{B}(\Xi)$ occurs in exactly one $\mathcal{B}((\Xi\cap H)\cup\zeta)$, the induction argument is complete. ◻

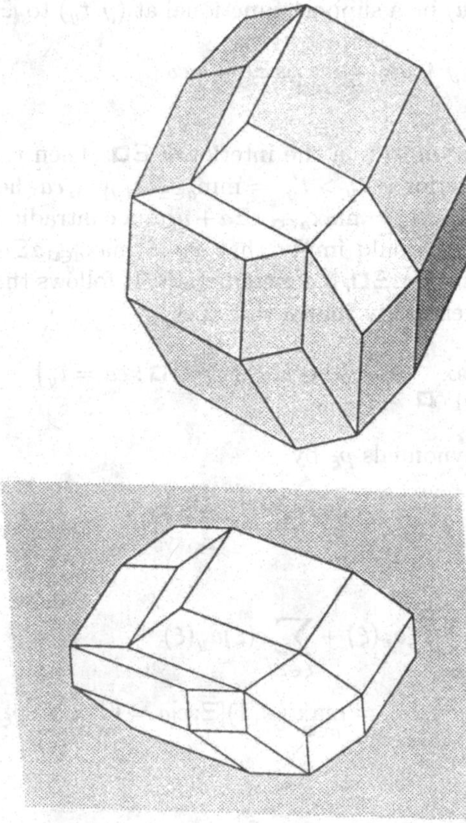

(57)Figure. A three-dimensional zonotope whose 'top' provides a tiling for the support of M_Ξ with $\Xi = \begin{bmatrix} 1 & 0 & 1 & -1 & -2 & 1 \\ 0 & 1 & 1 & 1 & 1 & 2 \end{bmatrix}$.

Zonotopes. Note that $\Xi\square$ is the sum of the 'intervals' $\xi[0\,..\,1]$, $\xi\in\Xi$. It is also known as the **zonotope** spanned by the columns of Ξ, because of the *belts* of parallelepipeds in the tilings of $\Xi\square$ described in (53)Theorem. In fact, it is possible to give a direct proof of (53)Theorem which, in effect, obtains such a tiling from the top or underside of the zonotope $[\Xi; c]\square \subset \mathbb{R}^{s+1}$, with $c \in \mathbb{R}^{\Xi}$ chosen 'generically', i.e., so that $x_Z := Z^{-T}(-c_{|Z}) \neq x_Y$ for all $Z \neq Y$ in $\mathcal{B}(\Xi)$. Here are the details.

Second proof of (53)Theorem. For each $y \in \Xi\square$, the point (y, t_y) with

$$t_y := \max_{a\in\Xi^{-1}\{y\}\cap\square} ca$$

is the highest point in $[\Xi; c]\square$ 'above' y. In particular, (y, t_y) is a boundary point of $[\Xi; c]\square$. Let (x, u) be a support functional at (y, t_y) to $[\Xi; c]\square$, i.e.,

$$xy + ut_y = \max_{a\in\square} x\Xi a + uca.$$

We now restrict attention to y in the interior of $\Xi\square$. Then u cannot be negative, since, for an interior y, $t_y > t'_y := \min_{a\in\Xi^{-1}\{y\}\cap\square} ca$, hence $u < 0$ would give $xy + ut'_y > xy + ut_y = \max_{a\in\square} x\Xi a + uca$, a contradiction. Also, u cannot be zero, for that would imply that $xy = \max_{a\in\square} x\Xi a$, hence y would be a boundary point for $\Xi\square$, a contradiction. It follows that we may assume without loss of generality that $u = 1$. Let

$$a_y \in \operatorname*{argmax}_{a\in\Xi^{-1}\{y\}\cap\square} ca = \{a \in \Xi^{-1}\{y\}\cap\square : ca = t_y\},$$

and define the linear polynomials p_ξ by

$$p_\xi : z \mapsto z\xi + c(\xi), \qquad \xi \in \Xi.$$

Then

$$\begin{aligned}\sum_{\xi\in\Xi} p_\xi(x)a_y(\xi) &= x\sum_{\xi\in\Xi}\xi a_y(\xi) + \sum_{\xi\in\Xi} c(\xi)a_y(\xi)\\ &= xy + ca_y = \max_{a\in\square}(x,1)[\Xi; c]a = \max_{a\in\square}\sum_{\xi\in\Xi} p_\xi(x)a(\xi),\end{aligned} \tag{58}$$

hence

$$a_y(\xi) \in \begin{cases}\{0\}, & p_\xi(x) < 0;\\ [0\,..\,1], & p_\xi(x) = 0;\\ \{1\}, & p_\xi(x) > 0.\end{cases} \tag{59}$$

We now show that we may choose x (for our given y) so that

$$\Xi_x := \{\xi \in \Xi : p_\xi(x) = 0\}$$

is in $\mathcal{B}(\Xi)$. Assume first that Ξ_x fails to be onto. Then there exists a nontrivial $z \perp \operatorname{ran}\Xi_x$. If $p_\xi(x+rz)p_\xi(x) > 0$ for all $r \in \mathbb{R}$ and all $\xi \notin \Xi_x$, then necessarily $z \perp \operatorname{ran}\Xi$, hence $\operatorname{ran}\Xi \neq \mathbb{R}^s$, contrary to assumption. Consequently, there exist r (necessarily not zero) and $\xi \notin \Xi_x$ so that $p_\xi(x+rz) = 0$. This implies that

$$r^* \in \operatorname{argmin}\{|r| : \exists\{\xi \notin \Xi_x\}\ p_\xi(x+rz) = 0\}$$

is not zero. Thus, $p_\xi(x+r^*z)p_\xi(x) \geq 0$ for all $\xi \in \Xi$, while $p_\xi(x+r^*z) = p_\xi(x) = 0$ for all $\xi \in \Xi_x$. Therefore, by (59), $\sum_{\xi\in\Xi} p_\xi(x+r^*z)(a_y - a)(\xi) \geq 0$ for all $a \in \square$, i.e., (58) holds with x replaced by $x+r^*z$, while Ξ_x is a proper subset of Ξ_{x+r^*z}. In other words, if Ξ_x is not onto, then we can enlarge it by changing to a different x.

It follows that we may assume without loss of generality that Ξ_x is onto. This implies that it contains some $\mathrm{Z} \in \mathcal{B}(\Xi)$, hence $x = x_\mathrm{Z} = \mathrm{Z}^{-T}(-c_{|\mathrm{Z}})$. If $\Xi_x \neq \mathrm{Z}$, then also $\mathrm{Z}' \subseteq \Xi_x$ for some $\mathrm{Z}' \in \mathcal{B}(\Xi)\backslash\mathrm{Z}$, and $x_\mathrm{Z} = x = x_{\mathrm{Z}'}$ would follow, a contradiction to our 'generic' choice of c. Note that such 'generic' choice is easily made inductively: with Y the collection of $\xi \in \Xi$ for which we have already chosen $c(\xi)$, we need, in choosing $c(\zeta)$ for the 'next' $\zeta \in \Xi\backslash Y$, only make certain that $-c(\zeta) \neq \zeta x_\mathrm{Z}$ for any $\mathrm{Z} \in \mathcal{B}(Y)$ (which excludes only finitely many choices for $c(\zeta)$).

We have shown that we may choose x so that $\Xi_x \in \mathcal{B}(\Xi)$. This implies, with (59), that $y \in \varphi(\mathrm{Z})$ for some $\mathrm{Z} \in \mathcal{B}(\Xi)$, where

$$\varphi(\mathrm{Z}) := \sum_{p_\xi(x_\mathrm{Z})>0} \xi + \mathrm{Z}\square.$$

Consequently, the interior of $\Xi\square$ lies in $\cup_{\mathrm{Z}\in\mathcal{B}(\Xi)}\varphi(\mathrm{Z})$. Since this is a finite union of closed sets, all in the closed set $\Xi\square$, it follows that

$$\Xi\square = \bigcup_{\mathrm{Z}\in\mathcal{B}(\Xi)} \varphi(\mathrm{Z}).$$

Finally, this union is essentially disjoint, for the following reason. If $y \in \varphi(\mathrm{Z})$, then $y = \Xi a_y$ for some a_y satisfying (59). Therefore $x_\mathrm{Z}y + ca_y = \max_{a\in\square}\sum_{\xi\in\Xi} p_\xi(x_\mathrm{Z})a(\xi)$. In particular,

$$\Xi_{a_y} := \{\xi \in \Xi : a_y(\xi) \in (0\,..\,1)\} \subseteq \Xi_{x_\mathrm{Z}}.$$

Thus, if also $y \in \varphi(\mathrm{Z}')$, then $\Xi_{a_y} \subseteq \Xi_{x_\mathrm{Z}} \cap \Xi_{x_{\mathrm{Z}'}}$. This implies that Ξ_{a_y} is not onto in case $\mathrm{Z} \neq \mathrm{Z}'$ (by our choice of c). We conclude that $\operatorname{vol}_s(\varphi(\mathrm{Z}) \cap \varphi(\mathrm{Z}')) = 0$ in case $\mathrm{Z} \neq \mathrm{Z}'$. ☐

The proof sets up a map from $\mathbb{R}^s$ to subsets of $\mathbb{R}^s$ by the rule

$$x \mapsto \sum_{p_\xi(x)>0} \xi + \Xi_x\square.$$

In this map, each connected component of the complement of the set $\cup_{\xi\in\Xi}\{y \in \mathbb{R}^s : p_\xi(y) = 0\}$ is mapped to a point (in $\Xi\square$) (since Ξ_x is empty for any x in that complement), with two such points differing by ξ exactly when the corresponding connected components share in their boundary an $(s-1)$-dimensional piece of the hyperplane $\{y \in \mathbb{R}^s : p_\xi(y) = 0\}$. Thus, by drawing in the straight line connecting any such pair, we obtain the tiling of $\Xi\square$. This is illustrated in (52)Figure.

Notes. The formal *definition* (1) first appeared in [de Boor, DeVore'83], as a generalization to arbitrary convex polytopes of Micchelli's definition [Micchelli'79] of Schoenberg's simplex spline (introduced in [de Boor'76h]). While M_Ξ was initially identified as having been obtained as the shadow of a parallelepiped, it was dubbed 'box spline' in [de Boor,Höllig'82b] (some would prefer the mathematically more dignified 'cube spline').

The realization that the hat function ([Courant'43]) can be thought of as the 'shadow' of a cube was the starting point for the use of box splines in [de Boor, DeVore'83] (and the end-of-proof symbol used in this book is meant as a related example). However, there are many other more challenging (bivariate) box splines that had been investigated well before 1982. We are indebted to J. Hoschek for the reference [Sommerfeld'04] which even includes a drawing of what we would now call a univariate B-spline, obtained there as the 'shadow' of a cube. The earliest references we know of in the context of multivariate splines are a paper by Di Guglielmo and a report by Frederickson. In [Guglielmo'70], the s-variate box spline with direction matrix $[\mathbb{1}_q, \mathbf{i}_1 + \cdots + \mathbf{i}_s]$ is introduced exactly as Schoenberg introduced the univariate cardinal B-spline in [Schoenberg'46a], namely via its Fourier transform. In [Frederickson'70] and its follow-up [Frederickson'71b], Frederickson uses convolution, by triangles, squares, and hat functions, to build up compactly supported smooth piecewise polynomials on a symmetric three-direction mesh. The same idea underlies the triangular splines of [Sabin'77]; see also [Sabin'89]. The Zwart-Powell element appears in [Zwart'73] as an example of that *rara avis*, a (nontrivial) compactly supported function constructed as a linear combination of 'half-space splines', i.e., functions of the form $x \mapsto (\xi x - c)_+^k$. It appears quite independently (apparently put together polynomial piece by polynomial piece) in [Powell'74]; see also [Powell, Sabin'77]. While working on [de Boor, DeVore'83], the authors were pleased to recognize that many of the triangular splines in [Sablonniere'81a] (not so called there) were special examples of a box spline; see also [Sablonniere'82a,'82b,'84a].

The *geometric description* (3) (with which [de Boor, DeVore'83] starts) recaptures the spirit of Schoenberg's definition of the simplex spline. The *inductive definition* (8) was already used in [Frederickson'70] as well as in [Sabin'77].

As to the notational *conventions*, the importance of $\mathcal{B}(\Xi)$ was first recognized in [Dahmen, Micchelli'85d] (see also [Dahmen, Micchelli'84d]), while the set $\mathcal{A}(\Xi)$ (alias $\mathcal{Y}(\Xi)$ in the papers by Dahmen & Micchelli) was already used (but not accorded a symbol) in [de Boor, Höllig'82b], in connection with the smoothness of a box spline. The early papers had no symbol for $\mathbb{H}(\Xi)$.

The various *basic properties* were first proved in [de Boor, Höllig'82b], with the following exceptions: The truncated power was introduced by Dahmen in [Dahmen'80c] (as the basic tool in his approach to simplex splines), and the remarkable formula (29) first appeared in [Dahmen, Micchelli'84f], while the fact that $D(\Xi) \subset \Pi$ was proved independently by Dahmen and Micchelli in [Dahmen, Micchelli'83b] using the Hilbert Nullstellensatz.

It is easy to mistake identities such as (20) or (22) for *pointwise* equalities. In fact, they are equations between distributions, hence hold pointwise only at points of continuity. It is, however, possible to follow Jia and *define* $M_\Xi(x)$ at a point x of discontinuity to be the limiting local average and, with such a pointwise definition of M_Ξ, equations such as (20) or (22) are indeed valid pointwise.

The *recurrence relations* also first appear in [de Boor, Höllig'82b], as a special case of the general recurrence relations for polyhedral splines in [de Boor, Höllig'82a]. However, the proof here is new. Computational experiences with the recurrence relations and alternative means for evaluating box splines are related in [de Boor'93].

We learned (53)Theorem concerning the *support of a box spline* from [Dahmen, Micchelli'84d,'85d(Theorem 3.3)] and made an effort to provide a complete, yet short, proof of this basic theorem. We learned only recently that this theorem can also be found in [Shepard'74], with the same inductive proof. The second proof, via *zonotopes*, was taken from a talk, by Andreas Dress at the Oberwolfach meeting 20-24 Oct 90, based on [Bohne, Dress, Fischer'89].

II

The linear algebra of box spline spaces

In this chapter, we consider the **cardinal spline space**

$$(1) \qquad S := S_M := \operatorname{span}\Big(M(\cdot - j)\Big)_{j\in\mathbb{Z}^s}$$

i.e., the space spanned by the **shifts**, or integer translates, of the box spline

$$M := M_\Xi.$$

We assume throughout that $\operatorname{ran}\Xi = \mathbb{R}^s$, hence that M_Ξ is a function on $\mathbb{R}^s$. Further, unless stated otherwise, we restrict Ξ to have *integer entries.*

Some of the discussion applies to an arbitrary compactly supported piecewise continuous function M. It will be clear from the context which results require specific box spline properties.

Since the index set

$$(2) \qquad \iota(x,\Xi) := (x - \Xi\square) \cap \mathbb{Z}^s \supseteq \{j \in \mathbb{Z}^s : M(x-j) \neq 0\}$$

is finite for any x, we feel free to consider *infinite* linear combinations of such shifts and interpret the resulting infinite sums

$$\sum_{j\in\mathbb{Z}^s} M(\cdot - j)a(j)$$

pointwise. In effect, we think of S as the *range* of the linear map $M^\star$, from **mesh functions** (i.e., functions on the integer lattice $\mathbb{Z}^s$) to functions on $\mathbb{R}^s$, given by

$$M^\star : a \mapsto M{*}a := \sum_{j\in\mathbb{Z}^s} M(\cdot - j)a(j). \tag{3}$$

Questions addressed in this chapter concern (i) the kernel of $M^\star$; hence (ii) on what sequence spaces $M^\star$ is 1-1, i.e., to what an extent the shifts of M are linearly independent; and (iii) what polynomials are contained in S. This last question is connected with the approximation power of the cardinal spline space.

Convolutions. In (3), we are using the familiar convolution notation in an unconventional but evident way to indicate convolution of a function with a sequence. The convention we use here and later is that $\varphi{*}a$ is given by

$$\varphi{*}a : x \mapsto \sum_{y\in \operatorname{dom}\varphi(x-\cdot)\cap \operatorname{dom} a} \varphi(x-y)a(y), \tag{4}$$

with the summation over the largest set on which both $\varphi(x-\cdot)$ and a are defined. In particular, $\varphi{*}a$ has the same domain and target as φ.

This convenient multi-use of the convolution star means that we cannot just write $M{*}$ for the map $f \mapsto M{*}f$ since we cannot guess from M alone what particular domain we have in mind for the map $f \mapsto M{*}f$. We reserve the notation $M{*}$ for the (standard) convolution map

$$M{*} : f \mapsto M{*}f := \int_{\mathbb{R}^s} M(\cdot - x)f(x)dx = \int_{\mathbb{R}^s} f(\cdot - x)M(x)dx$$

which carries functions on $\mathbb{R}^s$ to functions on $\mathbb{R}^s$. We use this notation even if M is not a function, but only a linear functional, i.e., when

$$M{*}f : x \mapsto Mf(x-\cdot).$$

For the two nonstandard convolution maps in this book, we use special symbols. We use the superscript five-pointed star * to indicate the map (3) from sequences to functions. We use the primed convolution star $*'$ for the **semi-discrete** convolution

$$M{*'} : f \mapsto M{*}(f_|) = \sum_{j\in\mathbb{Z}^s} M(\cdot - j)f(j)$$

of M with the *function* $f : \mathbb{R}^s \to \mathbb{R}$, i.e, the convolution of M with the sequence

$$f_| := f_{|\mathbb{Z}^s}.$$

We will make use of the fact that semi-discrete convolution commutes with shifts (i.e., integer translates), i.e., that, for any f and any $\alpha \in \mathbb{Z}^s$,

$$\begin{aligned}\tau_\alpha(M*'f) = (M*'f)(\cdot+\alpha) &= \sum_{j\in\mathbb{Z}^s} M(\cdot+\alpha-j)f(j)\\ &= \sum_{j\in\mathbb{Z}^s} M(\cdot-j)f(j+\alpha) = M*'\tau_\alpha f.\end{aligned} \tag{5}$$

Partition of unity. We show that the shifts of the box spline M form a partition of unity. By a standard argument (cf. Notes at chapter's end), this implies that the space spanned by these shifts can, at least, approximate continuous functions (as the mesh size is reduced by scaling). It also implies that $M*'$ is 1-1 on Π; cf. (14)Corollary.

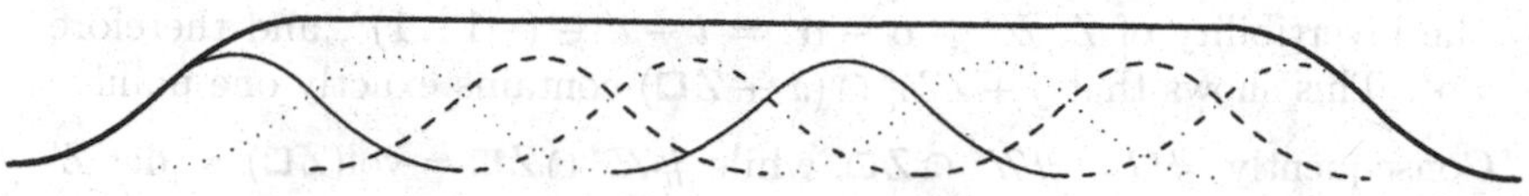

(6)Figure. The shifts of a box spline form a partition of unity.

We begin with the following technical lemma (see (12)Figure for an example with $x = 0$).

(7)Lemma. *For any invertible matrix* $Z \in \mathbb{Z}^{s\times s}$ *and any* $x \in \mathbb{R}^s$, $|\det Z| = \operatorname{vol} Z\square = \#\{\mathbb{Z}^s \cap (x + Z\square)\}$.

Proof. For any $r \in \mathbb{N}$, $r\square$ is the *disjoint* union $\{0,\ldots,r-1\}^s+\square$. As Z is invertible, this implies that $rZ\square$ is the disjoint union $Z\{0,\ldots,r-1\}^s+Z\square$, hence $\#\{(\mathbb{Z}^s - x)\cap rZ\square\} = r^s\#\{(\mathbb{Z}^s - x)\cap Z\square\}$. It follows that $\#(\mathbb{Z}^s\cap(x+Z\square)) = \#((\mathbb{Z}^s - x)\cap Z\square) = r^{-s}\#((\mathbb{Z}^s-x)\cap rZ\square) = r^{-s}\#((\mathbb{Z}^s - x)/r \cap Z\square)$, hence

$$\#(\mathbb{Z}^s\cap(x+Z\square)) = \sum_{y\in\left((\mathbb{Z}^s-x)/r\,\cap\, Z\square\right)} \operatorname{vol}(y+\square/r),$$

and this is a Riemann sum for the integral $\int_{Z\square} 1$, hence converges to $\operatorname{vol} Z\square$ as $r \to \infty$. $\square$

(8)Lemma. *For every* $\mathrm{Z} \in \mathcal{B}(\Xi)$,

$$\sum_{j\in\mathbb{Z}^s} M\Big(\cdot - \mathrm{Z}j\Big) = \frac{1}{|\det \mathrm{Z}|}.$$

Proof. Since $\mathbb{R}^s$ is the disjoint union of the (half-open) parallelepipeds $\mathrm{Z}(j+\square)$, $j \in \mathbb{Z}^s$, the lemma is valid for $\Xi = \mathrm{Z}$ by (I.9). The general case follows from this by repeated averaging, using (I.8) (or, equivalently, by (I.18)). □

(9)Lemma. *For every* $\mathrm{Z} \in \mathcal{B}(\Xi)$ *and every* $x \in \mathbb{R}^s$, *every coset* $j+\mathrm{Z}\mathbb{Z}^s$ *of the sublattice or subgroup* $A := \mathrm{Z}\mathbb{Z}^s$ *of* $\mathbb{Z}^s$ *meets the set* $x+\mathrm{Z}\square$ *in exactly one point. Consequently, its factor group* $G := \mathbb{Z}^s/A = \{j + A : j \in \mathbb{Z}^s\}$ *has finite order* $\#G = |\det \mathrm{Z}|$.

Proof. Since $x+\mathrm{Z}(\mathbb{Z}^s+\square) = \mathbb{R}^s$ (by the invertibility of Z), $j+\mathrm{Z}\mathbb{Z}^s$ must intersect $x + \mathrm{Z}(\beta+\square)$ for some β, hence, since $j + \mathrm{Z}\mathbb{Z}^s$ is invariant under translation by $-\mathrm{Z}\beta$, it must intersect $x + \mathrm{Z}\square$, i.e., there must exist $\alpha \in \mathbb{Z}^s$ and $t \in [0\,..\,1)^s$ so that $j + \mathrm{Z}\alpha = x + \mathrm{Z}t$. If also $j + \mathrm{Z}\alpha' = x + \mathrm{Z}t'$ for some $\alpha' \in \mathbb{Z}^s$ and some $t' \in [0\,..\,1)^s$, then $\mathrm{Z}(t-t') = \mathrm{Z}(\alpha-\alpha')$, hence, by the invertibility of Z, $\mathbb{Z}^s \ni \alpha - \alpha' = t - t' \in (-1\,..\,1)^s$, and therefore $\alpha = \alpha'$. This shows that $j + \mathrm{Z}\mathbb{Z}^s \cap (x + \mathrm{Z}\square)$ contains exactly one point.

Consequently, $\#G = \#\mathbb{Z}^s \cap \mathrm{Z}\square$, while $\#\mathbb{Z}^s \cap \mathrm{Z}\square = \mathrm{vol}(\mathrm{Z}\square) = |\det \mathrm{Z}|$ by (7)Lemma. □

For example, for $\Xi = \left[\begin{smallmatrix} 1 & 2 & 1 \\ -1 & 1 & 1 \end{smallmatrix}\right]$, the submatrix $\mathrm{Z} = \left[\begin{smallmatrix} 1 & 2 \\ -1 & 1 \end{smallmatrix}\right]$ is basic, with $\det \mathrm{Z} = 3$. (12)Figure shows the subgroup $A = \mathrm{Z}\mathbb{Z}^2$ as well as $\mathrm{Z}\square$, i.e., the (half-open) support of M_{Z}, and, in that support, a representer from each of the three cosets of A in $\mathbb{Z}^2$.

(10)Theorem. $\sum_{j\in\mathbb{Z}^s} M(\cdot - j) = 1$.

Proof. Pick any $\mathrm{Z} \in \mathcal{B}(\Xi)$. Then, setting $A := \mathrm{Z}\mathbb{Z}^s$ and $G := \mathbb{Z}^s/A$, we have from (8) and (9) that

$$\sum_{j\in\mathbb{Z}^s} M(\cdot - j) = \sum_{g\in G}\sum_{j\in A} M(\cdot - j - g) = \#G/|\det \mathrm{Z}| = 1.$$

□

(11)Corollary. *If* $\operatorname{ran}\Xi = \mathbb{R}^s$ *and* $|\det \mathrm{Z}| \neq 1$ *for some* $\mathrm{Z} \in \mathcal{B}(\Xi)$, *then* M^* *has (nontrivial) bounded sequences in its kernel.*

Proof. Under this assumption, (8)Lemma and (10)Theorem supply two different ways of writing the constant function 1 as a linear combination (with bounded coefficients) of box spline shifts. □

The converse of (11)Corollary is a consequence of (57)Theorem.

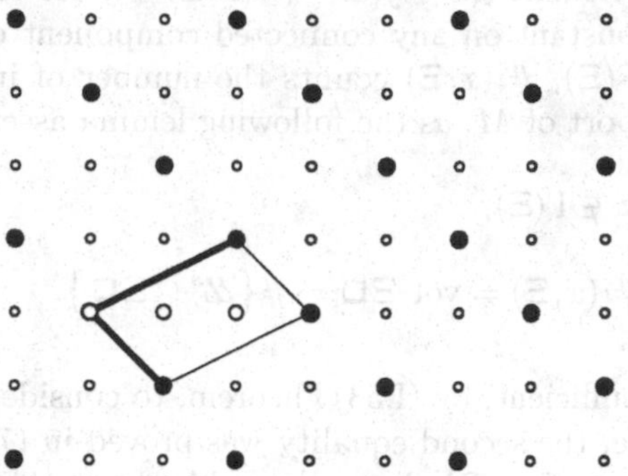

(12)Figure. The subgroup $\mathrm{Z}\mathbb{Z}^s$ of $\mathbb{Z}^s$ and the set $\mathrm{Z}\square$, for $\mathrm{Z} = \left[\begin{smallmatrix} 1 & 2 \\ -1 & 1 \end{smallmatrix}\right]$.

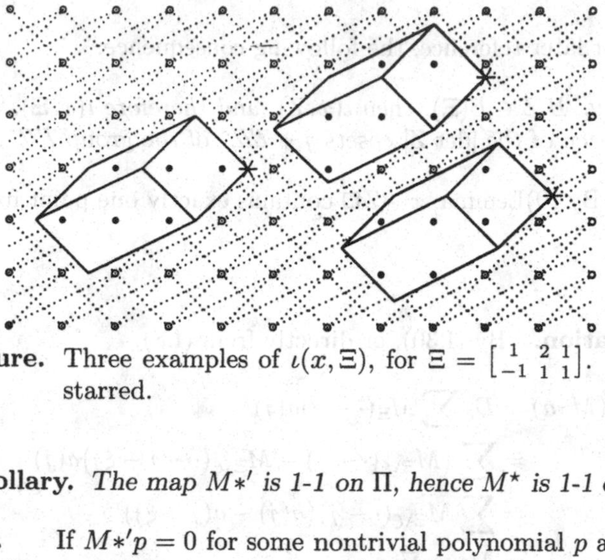

(13)Figure. Three examples of $\iota(x, \Xi)$, for $\Xi = \left[\begin{smallmatrix} 1 & 2 & 1 \\ -1 & 1 & 1 \end{smallmatrix}\right]$. The x's are starred.

(14)Corollary. *The map* $M*'$ *is 1-1 on* Π, *hence* $M^\star$ *is 1-1 on* $\Pi_|$.

Proof. If $M*'p = 0$ for some nontrivial polynomial p and therefore $r(\tau)p = 1$ for some difference operator $r(\tau)$, then

$$0 = r(\tau)M*'p = M*'r(\tau)p = M*'1$$

by (5), in contradiction to (10)Theorem. $\square$

The box spline shifts which do not vanish at a given point. Recall, from the discussion following (I.37)Proposition, the "mesh"

$$\Gamma(\Xi) = \bigcup\{H + \mathbb{Z}^s : H \in \mathbb{H}(\Xi)\}$$

generated by Ξ. For $x \notin \Gamma(\Xi)$, the set

$$\iota(x, \Xi) = \{j \in \mathbb{Z}^s : M_\Xi(x - j) \neq 0\} = \mathbb{Z}^s \cap (x - \Xi\square)$$

(introduced in (2)) provides the indices j of all box splines which do not vanish at x. In particular, $\iota(x+j,\Xi) = \iota(x,\Xi)+j$ for all $j \in \mathbb{Z}^s$. The function $\iota(\cdot,\Xi)$ is constant on any connected component of $\mathbb{R}^s \backslash \Gamma(\Xi)$. In fact, for each $x \notin \Gamma(\Xi)$, $\#\iota(x,\Xi)$ counts the number of integer points in the (half-open) support of M, as the following lemma asserts.

(15)Lemma. *For $x \notin \Gamma(\Xi)$,*

$$\#\iota(x,\Xi) = \operatorname{vol}\ \Xi\square = \#\{\mathbb{Z}^s \cap \Xi\square\}.$$

Proof. It is sufficient, by (I.53)Theorem, to consider the case when $\#\Xi = s$. In this case, the second equality was proved in (7)Lemma (recall our agreement that $\operatorname{ran}\Xi = \mathbb{R}^s$ throughout this chapter), as was the first, since $\#\iota(x,\Xi) = \#(\mathbb{Z}^s \cap (x - \Xi\square))$ for $x \notin \Gamma(\Xi)$ (and $\operatorname{vol}\ (-\Xi)\square = \operatorname{vol}\ \Xi\square$).

□

We note, for later reference, the following consequence:

(16)Corollary. *If $Z \in \mathcal{B}(\Xi)$, then $\iota(x,Z)$, and therefore $\iota(x,\Xi)$, contains a point from each of the $|\det Z|$ cosets $j + Z\mathbb{Z}^s$ of the group $Z\mathbb{Z}^s$.*

Proof. By (9)Lemma, $x - Z\square$ contains exactly one point from each coset $j + Z\mathbb{Z}^s$.

□

Differentiation. By (I.30), or directly from (I.8),

$$\begin{aligned} D_\xi(M{*}a) &= D_\xi \sum M_\Xi(\cdot - j)a(j) \\ &= \sum \big(M_{\Xi\backslash\xi}(\cdot - j) - M_{\Xi\backslash\xi}(\cdot - j - \xi)\big)a(j) \\ &= \sum M_{\Xi\backslash\xi}(\cdot - j)\big(a(j) - a(j-\xi)\big) \\ &= M_{\Xi\backslash\xi}{*}\nabla_\xi a, \end{aligned}$$

which yields the differentiation formula

$$(17) \qquad D_Z\big(M_\Xi{*}a\big) = M_{\Xi\backslash Z}{*}\nabla_Z a, \qquad \forall Z \subseteq \Xi.$$

Linear independence. By definition, the sequence $\big(M(\cdot - j)\big)_{j\in\mathbb{Z}^s}$ is (globally) **linearly independent** in case M^* is 1-1. The sequence is **locally linearly independent** if, for any bounded open G, all shifts of M having some support in G are linearly independent there. Remarkably, for a box spline M, these two notions of linear independence are equivalent (see (57)Theorem below).

Consideration of (global) linear independence leads to a study of $\ker M^\star$, i.e., a study of the collection of all sequences a for which $M{*}a = 0$. In the case of the box spline $M = M_\Xi$, the linear independence of the shifts of M_Ξ can already be characterized by considering $M_\Xi{}^\star$ on the sequence space

(18)
$$\Delta(\Xi) := \bigcap_{Z\in\mathcal{A}_{\min}(\Xi)} \ker \Delta_Z = \{f : \mathbb{Z}^s \to \mathbb{C} :\ \Delta_Z f = 0 \ \ \forall Z \in \mathcal{A}_{\min}(\Xi)\}.$$

By (17) (and since $\ker \Delta_\Xi = \ker \nabla_\Xi$),

(19)
$$M^\star(\Delta(\Xi)) \subseteq D(\Xi),$$

with $D(\Xi)$ the closely related *polynomial* space

(20)
$$D(\Xi) = \bigcap_{Z\in\mathcal{A}(\Xi)} \ker D_Z = \bigcap_{Z\in\mathcal{A}_{\min}(\Xi)} \ker D_Z \ \ \subseteq \Pi_k(\operatorname{ran}\Xi)$$

introduced in (I.37)Proposition (with $k = k(\Xi) = \#\Xi - d(\Xi)$); see also (59)Proposition.

Here, first in outline, is the beautiful theory of the various characterizations of the linear independence. We postpone proofs in order not to interrupt the narrative.

By (I.32) (with Ξ there replaced by $Z \in \mathcal{A}_{\min}(\Xi)$), $D(\Xi)_|$ is contained in $\Delta(\Xi)$, hence, by (19), $M{*}'(D(\Xi)) = M^\star(D(\Xi)_|) \subseteq M^\star(\Delta(\Xi)) \subseteq D(\Xi)$. Since $D(\Xi)$ is a finite-dimensional polynomial space by (20) and $M{*}'$ is 1-1 on Π by (14)Corollary, this implies that

(21)
$$M{*}'(D(\Xi)) = D(\Xi) = M^\star(\Delta(\Xi)) \subseteq S.$$

In general, $\Delta(\Xi)$ is larger than $D(\Xi)_|$ since, in addition to the polynomials from $D(\Xi)$, it may contain the restriction to $\mathbb{Z}^s$ of certain exponentials, as is to be expected from a system of difference equations. For example, for $\Xi = [3]$, we have $\mathcal{A}(\Xi) = \{[3]\}$, hence $D(\Xi) = \ker(3D) = \Pi_0$, but, since $\Delta_\Xi = \tau_3 - 1$, $\Delta(\Xi)$ also contains the *complex* exponentials $e_{\pm 1/3}$, with

(22)
$$e_z : \mathbb{Z}^s \to \mathbb{C} : \alpha \mapsto \exp(2\pi i z\alpha)$$

(with $z\alpha$ the scalar product). (30)Example provides a slightly more sophisticated (and relevant) illustration. As these examples show, it is convenient to admit also complex-valued functions, in order to capture easily all elements in the kernel of the difference operators of interest, as in the following (49)Theorem.

Theorem. *The sequence space* $\Delta(\Xi)$ *consists of discrete exponential polynomials. Precisely,*

$$\Delta(\Xi) = \oplus_{z \in \square} e_z D(\Xi_z)_{|}, \tag{23}$$

with

$$\Xi_z := \{\xi \in \Xi : \xi z \in \mathbb{Z}\}. \tag{24}$$

Here, for convenience of notation (yet entirely within the spirit of (20)), we set

$$D(\Xi_z) := \{0\} \ \text{ if } \ \operatorname{ran} \Xi_z \neq \mathbb{R}^s.$$

Thus, (23) has only finitely many (nontrivial) summands.

The direct summand in (23) corresponding to $z = 0$ is $D(\Xi)_{|}$ since $e_0 = 1$ and $\Xi_0 = \Xi$. We combine the remaining summands into one term,

$$E(\Xi) := \oplus_{z \in \square \backslash 0} e_z D(\Xi_z)_{|}, \tag{25}$$

and thus have

$$\Delta(\Xi) = D(\Xi)_{|} \oplus E(\Xi). \tag{26}$$

The following diagram summarizes a good part of the theory to be described:

$$\begin{array}{ccccccc} & & \mathbb{R}^{\iota(x,\Xi)} & & & & \\ & & \updownarrow \rho & & & & \\ D(\Xi)_{|} \oplus E(\Xi) & \longleftrightarrow & \Delta(\Xi) & \xrightarrow{M_\Xi{}^\star} & D(\Xi) & \rightarrow & 0 \end{array}$$

Explicitly, the map $M_\Xi{}^\star$ carries $\Delta(\Xi)$ onto $D(\Xi)$, by (21). Further, $M_\Xi{}^\star$ vanishes on $E(\Xi)$, hence $\ker M_\Xi{}^\star{}_{|\Delta(\Xi)} = E(\Xi)$ (this is (55)Theorem). Moreover, for any $x \notin \Gamma(\Xi)$, the map

$$\rho : \Delta(\Xi) \to \mathbb{R}^{\iota(x,\Xi)} : f \mapsto f_{|\iota(x,\Xi)} \tag{27}$$

is 1-1 and onto (this is (46)Theorem). In particular,

$$\dim \Delta(\Xi) = \#\iota(x,\Xi) = \operatorname{vol} \Xi\square, \tag{28}$$

the second equation by (15)Lemma, and $\Delta(\Xi)$ is uniquely determined by its restriction to $\iota(x,\Xi)$.

It is natural to consider $\iota(x,\Xi)$ in connection with *local* linear independence. For, the collection of all shifts of M_Ξ with some support in some sufficiently small G is precisely the set $\{M_\Xi(\cdot - j) : j \in \iota(x,\Xi)\}$ for some x (in G).

In addition, there is the following remarkable (32)Theorem:

Theorem. *For any* $\Xi \in \mathbb{R}^{s\times n}$ *with* $\operatorname{ran}\Xi = \mathbb{R}^s$, $\dim D(\Xi) = \#\mathcal{B}(\Xi)$.

These facts (still to be proved) provide most of the proof for the following crowning (57)Theorem.

Theorem. *Under the standing assumption on* Ξ *(i.e.,* $\Xi \in \mathbb{Z}^{s\times n}$, $\operatorname{ran}\Xi = \mathbb{R}^s$*), the following statements are equivalent.*

(i) *The shifts of* M_Ξ *are linearly independent.*

(ii) $M_\Xi{}^*$ *is 1-1 on* $\Delta(\Xi)$.

(iii) $\operatorname{vol}\Xi\square = \#\mathcal{B}(\Xi)$, *i.e.,* $\#\iota(x,\Xi) = \dim D(\Xi)$ *for all* $x \notin \Gamma(\Xi)$.

(iv) *The shifts of* M_Ξ *are locally linearly independent.*

(v) *All bases in* Ξ *have determinant* ± 1.

(vi) $E(\Xi) = 0$, *i.e.,* $\Delta(\Xi) = D(\Xi)_{|}$.

A matrix Ξ satisfying Condition (v) is said to be **unimodular**.

(29)Example: the three-direction mesh. If the shifts of the box spline M_Ξ are linearly independent, hence Ξ contains a basis with determinant ± 1 (by (57)(v)), then, after a change of variables (cf. (I.23)), we may assume (without giving up the assumption that Ξ is an integer matrix) that Ξ contains $\mathbb{1}$ as a submatrix. For $s = 2$, this further implies (with (57)(v)) that the columns of Ξ must come from the set

$$\{\pm\mathbf{i}_1,\ \pm\mathbf{i}_2,\ \pm\mathbf{i}_1 + \pm\mathbf{i}_2\},$$

with any two from the set $\{\pm\mathbf{i}_1 + \pm\mathbf{i}_2\}$ necessarily multiples of each other. Thus, by (I.20) and the symmetry in the variables, one can further assume that only the vectors $\mathbf{i}_1$, $\mathbf{i}_2$ and $\mathbf{i}_1 + \mathbf{i}_2$ occur. This is the reason for the prominence of the *three-direction mesh* in the study of *bivariate* box-splines. □

(30)Example: the ZP element. Consider the ZP element, i.e., the quadratic box spline corresponding to $\Xi = \left[\begin{smallmatrix}1&0&1&-1\\0&1&1&1\end{smallmatrix}\right]$. Since $\det\left[\begin{smallmatrix}1&-1\\1&1\end{smallmatrix}\right] = 2 \neq 1$, we know from (57)Theorem(v) that M^* is not 1-1. This is confirmed, via (57)(iii), by the fact that

$$\#\mathcal{B}(\Xi) = \binom{4}{2} = 6$$

since any two columns of Ξ span, while (cf. (I.47)Figure)

$$\operatorname{vol}\Xi\square = \sum_{Z\in\mathcal{B}(\Xi)} |\det Z| = 7.$$

By (21), $M_\Xi{}^\star$ carries $\Delta(\Xi)$ onto $D(\Xi)$, with kernel equal to $E(\Xi)$ (by (55)Theorem). Since $\dim \Delta(\Xi) = \operatorname{vol} \Xi \square$ by (28), while $\dim D(\Xi) = \#\mathcal{B}(\Xi)$ by (32)Theorem, this implies that $(\ker M^\star) \cap \Delta(\Xi) = E(\Xi)$ has dimension $\dim \Delta(\Xi) - \dim D(\Xi) = \operatorname{vol} \Xi\square - \#\mathcal{B}(\Xi) = 7 - 6 = 1$, hence any one nontrivial element of $E(\Xi)$ will span it. This is confirmed by the following: By (51)Remark, if Ξ_z is of full rank for some $z \in \square\backslash 0$, then necessarily $|\det \mathrm{Z}| \neq 1$ for every basis $\mathrm{Z} \subseteq \Xi_z$. Since any basis $\mathrm{Z} \subset \Xi$ containing one of the unit vectors necessarily has determinant ± 1, this shows that Ξ contains exactly one basis Ξ_z with $z \in [0\,.\,.\,1)^2\backslash 0$, namely $\Xi_z = \left[\begin{smallmatrix} 1 & -1 \\ 1 & 1 \end{smallmatrix}\right]$, corresponding to $z = (1/2, 1/2)$. Since this Ξ_z is of minimal order, $D(\Xi_z) = \Pi_0$, and $E(\Xi)$ is spanned by e_z, i.e., by the sequence

$$e_z(j) = \exp(2\pi i(j(1)/2 + j(2)/2)) = (-1)^{j(1)+j(2)}, \; j \in \mathbb{Z}^2.$$

□

The dimension of $D(\Xi)$. We know from (I.37)Proposition (and already used the fact, from (20)) that

$$D(\Xi) = \bigcap_{\mathrm{Z}\in\mathcal{A}(\Xi)} \ker D_\mathrm{Z} = \bigcap_{\mathrm{Z}\in\mathcal{A}_{\min}(\Xi)} \ker D_\mathrm{Z} \;\; \subseteq \Pi_k(\operatorname{ran}\Xi),$$

with $k = k(\Xi) = \#\Xi - d(\Xi)$.

In the *generic case*, i.e., when Ξ is in general position, we have in fact equality in (20), for the following reason: In the generic case, every square $\mathrm{Z} \subseteq \Xi$ is a basis (which is one way to *define* 'generic' here), hence $d = s$, and every $\mathrm{Z} \in \mathcal{A}(\Xi)$ has more than $k = n - s$ columns, and therefore $\Pi_k \subseteq D(\Xi)$.

(31)Proposition. *If Ξ is in general position, then $D(\Xi) = \Pi_k(\mathbb{R}^s)$.*

Since $\dim \Pi_k(\mathbb{R}^s) = \binom{k+s}{s} = \binom{n}{s} = \#\{\text{square } Z \subseteq \Xi\}$, it follows that, in the generic case, $\dim D(\Xi) = \#\mathcal{B}(\Xi)$. Remarkably, this connection between $\dim D(\Xi)$ and $\#\mathcal{B}(\Xi)$ persists in all cases. As the theorem to that effect deals with just the matrix Ξ, we relax our conditions on Ξ and allow Ξ to be an arbitrary matrix in $\mathbb{R}^{s\times n}$ with $\operatorname{ran}\Xi = \mathbb{R}^s$.

(32)Theorem. *For any $\Xi \in \mathbb{R}^{s\times n}$ with* $\operatorname{ran}\Xi = \mathbb{R}^s$, $\dim D(\Xi) = \#\mathcal{B}(\Xi)$.

Proof. The theorem is valid for $\#\Xi = s$, by (31)Proposition. Using induction, we assume that the theorem holds for all Ξ' with $\#\Xi' \leq \#\Xi$ and $\operatorname{ran}\Xi' = \mathbb{R}^s$, and establish it for the matrix $\Xi \cup \zeta$.

Recall that $\mathcal{A}_{\min}(\Xi) = \{\Xi\backslash H : H \in \mathbb{H}(\Xi)\}$, with $\mathbb{H}(\Xi)$ the collection of hyperplanes spanned by some columns of Ξ, and that $D(\Xi) \subseteq D(\Xi')$

in case $\Xi \subseteq \Xi'$ (by (I.41)Proposition), hence $D(\Xi) \subseteq D(\Xi \cup \zeta)$. Since $\Xi \backslash H \in \mathcal{A}_{\min}(\Xi \cup \zeta)$ if and only if $\zeta \in H$, it follows that

(33) $$D(\Xi) = D(\Xi \cup \zeta) \cap \bigcap_{\zeta \notin H} \ker D_{\Xi \backslash H}.$$

We proceed to construct, for each $\zeta \notin H \in \mathbb{H}(\Xi)$, an integral operator I_H so that

(i) I_H is 1-1 on Π;
(ii) $P_H : D(\Xi \cup \zeta) \to \Pi : f \mapsto I_H D_{\Xi \backslash H} f$ is a linear projector;
(iii) $\ker P_H = D(\Xi \cup \zeta) \cap \ker D_{\Xi \backslash H}$;
(34) (iv) $\operatorname{ran} P_H = I_H D(\Xi_H) \subseteq D(\Xi \cup \zeta)$, with
$$\Xi_H := (\Xi \cap H) \cup \zeta;$$
(v) $P_{H'} P_H = 0$ for $H' \neq H$.

From (ii,iv,v), we conclude that

$$P_0 := 1 - \sum_{H \in \mathbb{H}(\Xi)} P_H$$

is a linear projector on $D(\Xi \cup \zeta)$, and that

$$\operatorname{ran} P_0 = \bigcap_{\zeta \notin H} \ker P_H = D(\Xi \cup \zeta) \cap \big(\bigcap_{\zeta \notin H} \ker D_{\Xi \backslash H} \big) = D(\Xi),$$

by (iii) and (33). Since also $P_H P_0 = 0 = P_0 P_H$, it follows that $D(\Xi \cup \zeta)$ is the direct sum of the ranges of the projectors P_0 and P_H, $\zeta \notin H$, i.e.,

(35) $$D(\Xi \cup \zeta) = D(\Xi) \oplus \bigoplus_{\zeta \notin H} I_H D(\Xi_H).$$

Hence, from (i),

$$\dim D(\Xi \cup \zeta) = \dim D(\Xi) + \sum_{\zeta \notin H} \dim D(\Xi_H).$$

On the other hand, any basis in $\mathcal{B}(\Xi \cup \zeta)$ is either in $\mathcal{B}(\Xi)$ or is of the form $Z \cup \zeta$, with $Z \subset \Xi$ a basis for some hyperplane $H \in \mathbb{H}(\Xi)$ not containing ζ, and, for such an H, the latter bases are exactly the elements of $\mathcal{B}(\Xi_H)$. In other words,

$$\#\mathcal{B}(\Xi \cup \zeta) = \#\mathcal{B}(\Xi) + \sum_{\zeta \notin H} \#\mathcal{B}(\Xi_H).$$

This finishes the induction step since $\#\Xi_H \le \#\Xi$ and $\operatorname{ran}\Xi_H = \mathbb{R}^s$ imply that we have $\dim D(\Xi_H) = \#\mathcal{B}(\Xi_H)$ by induction.

The operator I_H is chosen to be the product of the integration operators I_ξ, $\xi \in \Xi\backslash H$, defined by

$$(I_\xi f)(z + t\xi) := \int_0^t f(z + w\xi)dw, \quad z \in H,$$

and this settles (34)(i). Since, in general, the operators I_ξ do not commute, we have to choose some definite order in the definition of I_H, but this order turns out not to matter. Since $D_\xi I_\xi = 1$ and $D_\xi D_\zeta = D_\zeta D_\xi$ for any ξ and ζ, we have

$$D_{\Xi\backslash H} I_H = 1$$

regardless of the order. Hence $I_H D_{\Xi\backslash H}$ is a linear projector, therefore so is its restriction P_H to $D(\Xi \cup \zeta)$. This settles (34)(ii,iii).

In the proof of (34)(iv,v), we use repeatedly that

$$D_Z I_H = D_{Z\backslash H} I_H D_{Z\cap H}$$

for any $Z \subseteq \Xi$, which is true since for each $z \in H$, D_z commutes with each of the I_ξ.

We start the proof of (34)(iv,v) by showing that

$$(36) \qquad I_H D(\Xi_H) \subseteq D(\Xi \cup \zeta).$$

For this, we need to show that $D_{(\Xi\cup\zeta)\backslash H'} I_H f = 0$ for every $H' \in \mathbb{H}(\Xi \cup \zeta)$ and every $f \in D(\Xi_H)$. There are two cases. If $H' = H$, this follows from the facts that $D_{\Xi\backslash H} I_H = 1$ and $[\zeta] \in \mathcal{A}_{\min}(\Xi_H)$. If $H' \ne H$, then

$$(37) \qquad H'_\zeta := (H \cap H') + \operatorname{ran}[\zeta] \;\in \mathbb{H}(\Xi_H) \subseteq \mathbb{H}(\Xi \cup \zeta).$$

Therefore

$$D_{(\Xi\cup\zeta)\backslash H'} I_H f = D_{(\dots)} I_H D_Z f,$$

with

$$\begin{aligned} Z &:= \Big((\Xi \cup \zeta) \cap H\Big)\backslash H' \supseteq \Big((\Xi \cap H) \cup \zeta\Big)\backslash\Big((H' \cap H) \cup \zeta\Big) \\ &= \Xi_H \backslash H'_\zeta \in \mathcal{A}_{\min}(\Xi_H), \end{aligned}$$

hence $D_{(\Xi\cup\zeta)\backslash H'} I_H f = 0$ in this case, too.

Since $(I_H D_{\Xi\backslash H}) I_H = I_H$ and $P_H = (I_H D_{\Xi\backslash H})_{|D(\Xi\cup\zeta)}$, (36) implies that

$$I_H D(\Xi_H) \subseteq P_H D(\Xi \cup \zeta).$$

For the proof of the converse containment, we observe that

$$D_{\Xi\backslash H}D(\Xi\cup\zeta)\subseteq D(\Xi_H), \tag{38}$$

by the first inclusion in (I.42) (with the substitution of $\Xi\cup\zeta$ for Ξ', and of $\Xi_H=(\Xi\cap H)\cup\zeta$ for Ξ, hence $\Xi'\backslash\Xi$ becomes $\Xi\backslash H$). Thus (applying I_H to both sides of (38)),

$$P_H D(\Xi\cup\zeta)\subseteq I_H D(\Xi_H).$$

This proves that

$$P_H D(\Xi\cup\zeta)=I_H D(\Xi_H)\subseteq D(\Xi\cup\zeta),$$

which is (34)(iv).

Finally, for the proof of (34)(v), we note that, on $D(\Xi\cup\zeta)$,

$$P_{H'}P_H=I_{H'}D_{\Xi\backslash H'}I_H D_{\Xi\backslash H}=I_{H'}D_{(\ldots)}I_H D_{\mathrm{Z}},$$

with

$$\begin{aligned}\mathrm{Z}:=\Big((\Xi\backslash H')\cap H\Big)\cup\Big(\Xi\backslash H\Big)&=\Xi\backslash(H'\cap H)\\ &=(\Xi\cup\zeta)\backslash((H'\cap H)\cup\zeta)\ \in\mathcal{A}_{\min}(\Xi\cup\zeta),\end{aligned}$$

the containment by (37). □

(39) Remark. It is the *in*equality $\dim D(\Xi)\ge\#\mathcal{B}(\Xi)$ which makes the proof just given as long as it is. The converse inequality $\dim D(\Xi)\le\#\mathcal{B}(\Xi)$ is much easier to prove since it merely requires a proof of the inequality

$$\dim D(\Xi\cup\zeta)\le\dim D(\Xi)+\sum_{\zeta\notin H}\dim D(\Xi_H),$$

and this can be obtained from the basic linear algebra observation that $\dim\operatorname{dom}A=\dim\ker A+\dim\operatorname{ran}A$ by considering the particular linear map

$$A:D(\Xi\cup\zeta)\to\mathop{\times}_{\zeta\notin H}D(\Xi_H):f\mapsto\big(D_{\Xi\backslash H}f\big)_{\zeta\notin H}.$$

(40) Remark. If the columns of Ξ are in general position (i.e. if every $\mathrm{Z}\subseteq\Xi$ with $\#\mathrm{Z}=s$ spans), then, as mentioned before, $\#\mathcal{B}(\Xi)=\binom{n}{s}$ and $D(\Xi)=\Pi_k$. More generally, denote by Ξ_r the matrix made up from the columns of Ξ, with $\xi\in\Xi$ repeated with multiplicity $r(\xi)$. Then, for Ξ in general position,

$$\#\mathcal{B}(\Xi_r)=\sum_{\mathrm{Z}\in\mathcal{B}(\Xi)}\prod_{\xi\in\mathrm{Z}}r(\xi). \tag{41}$$

In particular, if $r(\xi) = r$ for all ξ, then $\#\mathcal{B}(\Xi_r) = \binom{n}{s} r^s$.

To illustrate the assertion of (32)Theorem, we discuss two special cases.

(42)Tensor products. By this we mean that, for some $r \in \mathbb{N}^s$,

$$\Xi = \mathbb{1}_r = [\ldots, \underbrace{\mathbf{i}_i, \ldots, \mathbf{i}_i}_{r(i) \text{ terms}}, \ldots].$$

Then, by (41), $\#\mathcal{B}(\Xi) = \prod_{i=1}^s r(i)$. On the other hand,

$$D(\Xi) = \bigcap_{i=1}^{s} \ker D_i^{r(i)} = \Pi_{(r(1)-1,\ldots,r(s)-1)},$$

which confirms (32)Theorem in this case.

More generally, if $\Xi = Z_r$ with Z a basis, then

$$D(\Xi) = \operatorname{span}\{\prod_{\nu=1}^{s} (\cdot\eta_\nu)^{\alpha(\nu)} : \alpha(\nu) < r(\nu)\ \forall\nu\},$$

with η_ν the νth row of Z^{-1}, hence $D_\xi(\cdot\eta_\nu)^\mu = \mu(\xi\eta_\nu)(\cdot\eta_\nu)^{\mu-1}$.

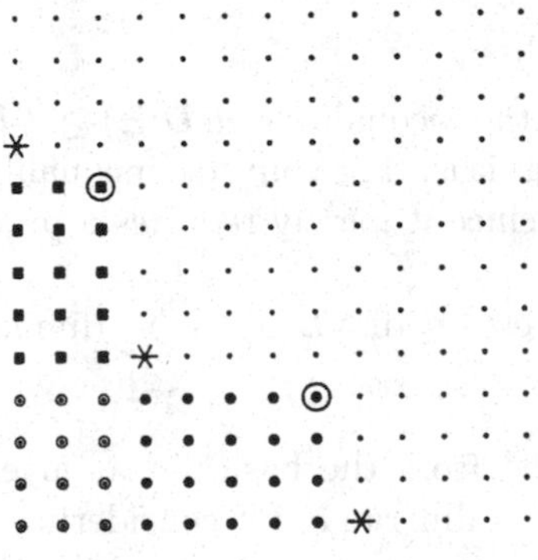

(43)Figure. Schematic description of $D(\Xi)$ (for $\Xi = [\mathbf{i}_1, \mathbf{i}_2, \mathbf{i}_1 + \mathbf{i}_2]_{(3,4,5)}$) showing a basis, a minimal generating set (circled) for $D(\Xi)$ and a complete set of generators (starred) of its annihilator.

(44)Three-direction mesh. The simplest mesh which is not of tensor product type is generated by the three vectors $\xi := \mathbf{i}_1 = (1,0)$, $\eta := \mathbf{i}_2 = (0,1)$, $\zeta := \xi + \eta$, i.e.,

$$\Xi := T_r := \begin{bmatrix} 1 & 0 & 1 \\ 0 & 1 & 1 \end{bmatrix}_r,$$

with $r =: (\alpha, \beta, \gamma)$ specifying the multiplicity of ξ, η and ζ in Ξ. In particular, $\gamma = 0$ corresponds to the tensor product case.

To simplify notation, we refer to T_r just by the multiplicities r, e.g., we write $D(T_r) =: D(r)$. By (41), we have

$$\dim D(r) = \#\mathcal{B}(r) = \alpha\beta + \alpha\gamma + \beta\gamma.$$

Following the inductive argument in the proof of (32)Theorem, we construct a basis for $D(r)$. We begin by recalling that, for the tensor product case,

$$D(\alpha, \beta, 0) = \Pi_{\alpha-1,\beta-1}$$

and $\dim D(\alpha, \beta, 0) = \alpha\beta$. Then we apply (35) repeatedly with $\zeta = (1,1)$. Since the hyperplanes H in $\mathbb{H}(\Xi)$ with $\zeta \notin H$ are the lines $\xi\mathbb{R}$, $\eta\mathbb{R}$, this gives

$$D(\alpha, \beta, \nu+1) = D(\alpha, \beta, \nu) \oplus I(0, \beta, \nu)D(\alpha, 0, 1) \oplus I(\alpha, 0, \nu)D(0, \beta, 1)$$

with $I(a, b, c) := I_\xi^a I_\eta^b I_\zeta^c$ the product of integral operators. Thus,

$$\begin{aligned} D(\alpha, \beta, \gamma) &= D(\alpha, \beta, 0) \oplus \bigoplus_{\nu<\gamma} I(0, \beta, \nu)D(\alpha, 0, 1) \oplus \bigoplus_{\nu<\gamma} I(\alpha, 0, \nu)D(0, \beta, 1) \\ &=: K(\zeta) \oplus K(\xi) \oplus K(\eta). \end{aligned}$$

An explicit basis for $K(\xi)$ can be constructed by integrating the basis

$$[\![\zeta^\perp \cdot]\!]^\mu, \quad 0 \le \mu < \alpha,$$

for $D(\alpha, 0, 1)$, with $\zeta^\perp := (1, -1)$. For this observe that, e.g.,

$$(I_\zeta^\nu f)(z + t\zeta) = \int_0^t [\![t - w]\!]^{\nu-1} f(z + w\zeta)dw, \qquad z \in \xi\mathbb{R}.$$

Hence, with $(u, v) = (u - v)\xi + v\zeta$,

$$I(0,0,\nu)[\![\zeta^\perp \cdot]\!]^\mu(u,v) = \int_0^v [\![v-w]\!]^{\nu-1} [\![\zeta^\perp((u-v)\xi + w\zeta)]\!]^\mu dw = [\![u-v, v]\!]^{\mu,\nu}$$

and so

$$b_{\mu,\beta+\nu} := I(0, \beta, \nu)[\![\zeta^\perp \cdot]\!]^\mu : (u,v) \mapsto \int_0^v [\![v - w]\!]^{\beta-1} [\![u - w, w]\!]^{\mu,\nu} dw,$$

$$\mu < \alpha, \ \nu < \gamma,$$

provides a basis for $K(\xi)$. By symmetry, the basis $b_{\alpha+\nu,\mu}$, $\nu < \gamma$, $\mu < \beta$, for $K(\eta)$ is obtained by interchanging u and v. Setting $b_j := [\![]\!]^j$, $j \le (\alpha - 1, \beta - 1)$, completes the basis for $D(\Xi)$. Only these latter basis elements are actually monomials, but all b_j so constructed are homogeneous polynomials, of total degree $|j|$. This makes it reasonable to visualize them as in (43)Figure, in which b_j is indicated by a symbol at mesh point j, with different symbols depending on whether b_j belongs to $K(\zeta), K(\xi)$, or $K(\eta)$. In this way, the number of marks in the initial triangle $|j| \le k$ equals the dimension of $D(\Xi) \cap \Pi_k$. The figure also marks the two **generators** of $D(\Xi)$, i.e., two polynomials $b_{\alpha+\gamma-1,\beta-1}$ and $b_{\alpha-1,\beta+\gamma-1}$ which, together with their derivatives, span $D(\Xi)$. Finally, the figure also marks the three **generators** of the set of differential operators whose joint kernel is $D(\Xi)$, i.e., the three differential operators $D^{\alpha+\gamma,0}, D^{\alpha,\beta}, D^{0,\beta+\gamma}$. The resulting image of interlocking corners is qualitatively correct for arbitrary Ξ with $\operatorname{ran}\Xi = \mathbb{R}^2$. See (65)Remark for a discussion of the fact that (43) is qualitatively correct for general Ξ.

The structure of $\Delta(\Xi)$. We are now ready to prove the specific statements about $\Delta(\Xi)$ mentioned earlier. Recall from (24) the definition

$$\Xi_z := \{\xi \in \Xi : \xi z \in \mathbb{Z}\}$$

and the agreement that $D(\Xi_z) = \{0\}$ in case Ξ_z fails to span.

(45)Lemma. *For any $z \in \mathbb{C}^s$, $e_z D(\Xi_z)_| \subseteq \Delta(\Xi)$.*

Proof. Let f be any sequence of the form $f := e_z p_|$, with $p \in D(\Xi_z)$, and let $H \in \mathbb{H}(\Xi)$. Since $\Delta_\xi(e_z p_|) = e_z(\Delta_\xi p_|)$ for any $\xi \in \Xi_z$, we have $\Delta_Z(e_z p_|) = e_z \Delta_Z p_|$ for $Z := (\Xi \backslash H) \cap \Xi_z$. As such Z is necessarily in $\mathcal{A}(\Xi_z)$, this implies that

$$\Delta_{\Xi\backslash H} e_z p_| = \Delta_{(\dots)}(e_z \Delta_Z p_|) = 0,$$

since $p_| \in D(\Xi_z)_| \subseteq \Delta(\Xi_z)$, by (I.32). □

(46)Theorem. *For $x \notin \Gamma(\Xi)$, the restriction map*

$$\rho : \Delta(\Xi) \to \mathbb{R}^{\iota(x,\Xi)} : f \mapsto f_{|\iota(x,\Xi)}$$

is 1-1 and onto.

This theorem implies that any mesh function on $\iota(x,\Xi)$ (with $x \notin \Gamma(\Xi)$) can be (uniquely) extended to an element of $\Delta(\Xi)$. One may consider, more generally, mesh functions defined on

$$\iota(\Omega, \Xi) := \cup_{x \in \Omega} \iota(x, \Xi).$$

There is no reason to believe that arbitrary mesh functions on $\iota(\Omega,\Xi)$ can be extended to an element of $\Delta(\Xi)$. But, as is shown in (VI.38)Proposition, any mesh function f on $\iota(\Omega,\Xi)$ for which $\Delta_Z f = 0$ for all $Z \in \mathcal{A}_{\min}(\Xi)$, to the extent that this makes sense for an f only defined on $\iota(\Omega,\Xi)$, can indeed be extended to an element of $\Delta(\Xi)$, in case Ω is connected.

Proof. We start the proof by showing that ρ is 1-1, i.e., that

$$\ker\rho = 0.$$

We proceed by induction on $\#\Xi$. Let $f \in \Delta(\Xi)$ vanish on $\iota(x,\Xi)$. We claim that, for any $\xi \in \Xi$, $\Delta_\xi f = 0$.

There are two cases. If $\Xi\backslash\xi$ fails to span, then $\Delta_\xi f = 0$ for any $f \in \Delta(\Xi)$. If, on the other hand, $\Xi\backslash\xi$ spans, then $\Delta_\xi f \in \Delta(\Xi\backslash\xi)$, and, for every $j \in \iota(x,\Xi\backslash\xi)$, both j and $j-\xi$ are in $\iota(x,\Xi)$, hence $\Delta_\xi f(\cdot - \xi)$ vanishes on $\iota(x,\Xi\backslash\xi)$, and must therefore vanish identically, by induction; hence so must $\Delta_\xi f$.

In particular, $\Delta_\zeta f = 0$ for all ζ in any particular basis $\mathrm{Z} \in \mathcal{B}(\Xi)$. This implies that f is constant on all cosets $j + \mathrm{Z}\mathbb{Z}^s$. By (16)Corollary, each of these cosets intersects $\iota(x,\Xi)$. Hence $f = 0$.

With this,

(47)
$$\dim\Delta(\Xi) = \dim\ker\rho + \dim\operatorname{ran}\rho = 0 + \dim\operatorname{ran}\rho \le \#\iota(x,\Xi) = \operatorname{vol}\ \Xi\square,$$

(the last equation by (15)Lemma). Hence it follows that ρ is onto once we show that

(48)
$$\dim\Delta(\Xi) \ge \operatorname{vol}\ \Xi\square$$

which we do in the proof of the next theorem. □

(49)Theorem. *The sequence space* $\Delta(\Xi)$ *consists of discrete exponential polynomials. Precisely,*

(50)
$$\Delta(\Xi) = \oplus_{z\in\square} e_z D(\Xi_z)_|,$$

with

$$\Xi_z = \{\xi \in \Xi : \xi z \in \mathbb{Z}\}.$$

As a consequence, $\dim\Delta(\Xi) = \operatorname{vol}\ \Xi\square$.

Proof. We know from (45)Lemma that $\Delta(\Xi) \supseteq \sum_{z\in\square} e_z D(\Xi_z)_|$. Since this sum is over $z \in \square$, it is direct (as is proved below, for completeness, in (52)Corollary to (53)Lemma). Therefore,

$$\dim\Delta(\Xi) \ge \sum_{z\in\square} \dim D(\Xi_z) = \sum_{\substack{z\in\square \\ \operatorname{ran}\Xi_z = \mathbb{R}^s}} \#\mathcal{B}(\Xi_z).$$

(The restriction to $\operatorname{ran}\Xi_z = \mathbb{R}^s$ is needed since $\mathcal{B}(\Xi_z)$ is, by definition, the collection of bases for $\operatorname{ran}\Xi_z$.) By definition of Ξ_z, the right hand side equals

$$\sum_{z\in\square}\#\{Z\in\mathcal{B}(\Xi) : Z^T z\in\mathbb{Z}^s\} = \sum_{Z\in\mathcal{B}(\Xi)}\#\{z\in\square : Z^T z\in\mathbb{Z}^s\}$$
$$= \sum_{Z\in\mathcal{B}(\Xi)}|\det Z| = \operatorname{vol}\,\Xi\square,$$

the second equality by (7)Lemma and the last by (I.53)Theorem. This proves (48) and so finishes the proof of (46)Theorem.

In particular, we now know that equality must hold in (47) and (48), and this completes the proof of the present theorem. □

(51)Remark. If $D(\Xi_z)\neq 0$ (i.e. if $\operatorname{ran}\Xi_z = \mathbb{R}^s$) with $z\in\square\backslash 0$, then necessarily

$$|\det Z|\neq 1 \quad \text{for every basis } Z\subseteq\Xi_z.$$

For, if $|\det Z| = 1$, then, by Cramer's rule, the linear system $Z^T z = j$ with $j\in\mathbb{Z}^s$ has a unique integer solution, hence $z\in\square$ implies that $z=0$.

For completeness, we now provide the proof that the sum $\sum_z e_z P_z$, with P_z arbitrary polynomial spaces but with $z\in\square$, is direct, with the aid of the following lemma.

(52)Lemma. *For any k, any finite set $\mathcal{Z}\subseteq\square$ and any $w\in\mathcal{Z}$, there exists a difference operator $r(\tau)$ which is 1-1 on $e_w\Pi$ and annihilates every $e_z\Pi_{<k}$ with $z\in\mathcal{Z}\backslash w$.*

Proof. One calculates that, for arbitrary $y\in\mathbb{R}^s$, $c\in\mathbb{C}$, $z\in\mathbb{C}^s$, and $p\in\Pi$,

$$(\Delta_y - c)(e_z p) = e_z\big((e_z(y) - 1 - c)p + e_z(y)\Delta_y p\big) =: e_z q.$$

This shows that the difference operator $\Delta_y - c$ is **degree-preserving** in the sense that $(\Delta_y - c)(e_z p) = e_z q$ with $\deg q = \deg p$, provided $e_z(y)-1-c\neq 0$. In the contrary case, $\deg q < \deg p$, hence $(\Delta_y - c)^k$ annihilates every $e_z p$ with $\deg p < k$. In particular, for any pair z, w with $z - w\notin\mathbb{Z}^s$, we can choose $y\in\mathbb{Z}^s$ and $c\in\mathbb{C}$ so that $(\Delta_y - c)^k$ vanishes on $e_z\Pi_{<k}$, yet is degree-preserving on $e_w\Pi$. Further, by restricting y to lie in $\mathbb{Z}^s$, we can be certain that $\Delta_y - c$ is a difference operator, i.e., of the form $r(\tau)$ for some polynomial r.

Since $\mathcal{Z}\subset\square$, the difference $w - z$ of any two different $w, z\in\mathcal{Z}$ is never in $\mathbb{Z}^s$. As there are only finitely many $z\in\mathcal{Z}$, we can therefore find, for any k and any $w\in\mathcal{Z}$, a difference operator of the desired kind as a product of operators of the above kind. □

(53)Corollary. *If $f := \sum_{z\in\mathcal{Z}} e_z p_z = 0$ on $\mathbb{Z}^s$ for some finite set $\mathcal{Z} \subseteq \square$, with $p_z \in \Pi$ for all z, then $p_z = 0$ for all z.*

Proof. Since $f_| = 0$, the difference operator $r(\tau)$ constructed in the preceding lemma for $w \in \mathcal{Z}$ gives $0 = r(\tau)f_| = r(\tau)e_w p_{w|} = e_w q_|$, hence $q_| = 0$, for some polynomial q with $\deg q = \deg p_w$. Consequently, $\deg p_w < 0$, i.e., $p_w = 0$. ☐

(54)Remark. The same argument supports the analogous claim that the vanishing on all of $\mathbb{R}^s$ of the finite sum $\sum_{z\in\mathcal{Z}} e_z p_z$, with $p_z \in \Pi$ and $\mathcal{Z} \subset \mathbb{R}^s$, implies that $p_z = 0$ for all $z \in \mathcal{Z}$.

(55)Theorem. $\ker M_\Xi{}^\star{}_{|\Delta(\Xi)} = E(\Xi)$.

Proof. Since $M_\Xi{}^\star$ is 1-1 on $D(\Xi)_|$ (by (20) and (14)Corollary), the kernel of $M_\Xi{}^\star{}_{|\Delta(\Xi)}$ lies in some algebraic complement of $D(\Xi)_|$ in $\Delta(\Xi)$. Since $\Delta(\Xi) = D(\Xi)_| \oplus E(\Xi)$ by (26), it is therefore sufficient to show that $E(\Xi)$ lies in the kernel of $M_\Xi{}^\star$, i.e., that

$$q := M_\Xi * (e_z p_|) = 0, \quad \text{for } p \in D(\Xi_z),\ z \in \square\backslash 0. \tag{56}$$

By (19) and (20), q is a polynomial. Consequently, by (53)Lemma (with $w = 0$ and $\mathcal{Z} = \{0, z\}$), we can find a difference operator $r(\tau)$ which annihilates $e_z p$ and maps q to a polynomial of the same degree as q. Since (with (5)) $r(\tau)q = r(\tau)(M_\Xi * (e_z p_|)) = M_\Xi * (r(\tau)e_z p_|) = M_\Xi * 0 = 0$, this implies that $q = 0$. ☐

(57)Theorem. *Under the standing assumptions on Ξ (i.e., $\Xi \in \mathbb{Z}^{s\times n}$, $\operatorname{ran}\Xi = \mathbb{R}^s$), the following statements are equivalent.*

(i) *The shifts of M_Ξ are linearly independent, i.e., $M_\Xi{}^\star$ is 1-1.*

(ii) *$M_\Xi{}^\star$ is 1-1 on $\Delta(\Xi)$.*

(iii) $\operatorname{vol}\Xi\square = \#\mathcal{B}(\Xi)$, *i.e.,* $\#\iota(x, \Xi) = \dim D(\Xi)$ *for all* $x \notin \Gamma(\Xi)$.

(iv) *The shifts of M_Ξ are locally linearly independent.*

(v) *All bases in Ξ have determinant ± 1 (i.e., Ξ is unimodular).*

(vi) $E(\Xi) = 0$, *i.e.,* $\Delta(\Xi) = D(\Xi)_|$.

Proof. The equivalence of the two conditions in (iii) follows from (28) and from (32)Theorem. The equivalence of the two conditions in (vi) follows from (26).

(i) $\Longrightarrow$ (ii): Trivial.

(ii) $\Longrightarrow$ (iii): The map $\rho : \Delta(\Xi) \to \mathbb{R}^{\iota(x,\Xi)} : f \mapsto f_{|\iota(x,\Xi)}$ is invertible (by (46)Theorem), and $M_\Xi{}^\star$ maps $\Delta(\Xi)$ onto $D(\Xi)$ (by (21)).

(iii) $\Longrightarrow$ (iv): $\#\iota(x,\Xi)$ counts the shifts of M_Ξ which are not zero on the connected component G of $\mathbb{R}^s\backslash\Gamma(\Xi)$ containing x, and these necessarily span $D(\Xi)_{|G}$ by (21), and $\dim D(\Xi)_{|G} = \dim D(\Xi)$ (since $D(\Xi)$ is a polynomial space).

(iv) $\Longrightarrow$ (i): Local linear independence is stronger than linear independence.

(v) $\Longleftrightarrow$ (iii): by (I.53)Theorem.

(vi) $\Longleftrightarrow$ (iii): By (26), (vi) is equivalent to having $\dim \Delta(\Xi) = \dim D(\Xi)$ thus equivalent to (iii) by (49)Theorem and (32)Theorem. ❏

(58)Example: four-direction mesh with multiplicities. We continue (30)Example by considering, more generally, the matrix Ξ_r, where $r(\nu)$ denotes the multiplicity of the νth direction, ξ_ν, in Ξ, as in (40), and $\Xi = \begin{bmatrix} 1 & 0 & 1 & -1 \\ 0 & 1 & 1 & 1 \end{bmatrix}$ is the matrix associated with the ZP element. In this case,

$$\#\mathcal{B}(\Xi_r) = \sum_{\nu<\mu} r(\nu)r(\mu)$$

by (41), while

$$\operatorname{vol} \Xi\square = \sum_{\nu<\mu} |\det[\xi_\nu, \xi_\mu]| r(\nu)r(\mu) = \#\mathcal{B}(\Xi_r) + r(3)r(4).$$

As already explained in (30)Example, a spanning $\Xi_z \subseteq \Xi_r$ with $z \in \square\backslash 0$ can only contain the vectors $(1,1)$ and $(-1,1)$. Therefore,

$$\Xi_z = \begin{bmatrix} 1 & -1 \\ 1 & 1 \end{bmatrix}_{(r(3),r(4))}$$

and $z = (1/2, 1/2)$. The sequences in $E(\Xi) = e_z D(\Xi_z)$ are of the form (cf. (42)Example)

$$(-1)^{|j|} \sum_{\alpha \le (r(3)-1, r(4)-1)} a_\alpha\, (j\xi_3)^{\alpha(1)} (j\xi_4)^{\alpha(2)}, \quad j \in \mathbb{Z}^2.$$

❏

This finishes the discussion of linear independence of box spline shifts. We turn now to the second topic of concern, namely the polynomials contained in the cardinal spline space S_M.

The polynomials contained in S. Recall that $S := S_M$ is, by definition, the space spanned by all shifts of the compactly supported piecewise continuous function M, specifically, $M = M_\Xi$. We are interested in the space

$$\Pi_M := \Pi \cap S$$

of all polynomials in S.

(59)Proposition. *The space Π_M of polynomials in the cardinal spline space $S = S_M$ generated by the box spline $M = M_\Xi$ with $\operatorname{ran}\Xi = \mathbb{R}^s$ coincides with $D(\Xi)$. More precisely,*

$$\Pi_{m(\Xi)} \subseteq \Pi_M = D(\Xi) \subseteq \Pi_{k(\Xi)}, \tag{60}$$

and these inclusions are sharp in the sense that $\Pi_\mu \subseteq D(\Xi) \subseteq \Pi_\kappa$ implies that $\mu \le m(\Xi)$ and $k(\Xi) \le \kappa$. In particular, there is any equality if and only if $m(\Xi) = k(\Xi)$, i.e., if and only if Ξ is in general position.

Proof. Recall from (I.37)Proposition that the box spline $M = M_\Xi$ is piecewise in

$$D(\Xi) = \bigcap_{Z \in \mathcal{A}(\Xi)} \ker D_Z \subseteq \Pi_{k(\Xi)}, \tag{61}$$

hence $\Pi_M = \Pi \cap S \subseteq D(\Xi)$ in this case, while the reverse inclusion $D(\Xi) \subseteq \Pi \cap S = \Pi_M$ follows from (21).

As for the first inclusion in (60), note that all differential operators appearing in (61) have order at least

$$\min\{\#Z : Z \in \mathcal{A}(\Xi)\} = m(\Xi) + 1,$$

hence $\Pi_{m(\Xi)} \subseteq D(\Xi)$. To verify the sharpness of this inclusion, observe that, by definition of $m(\Xi)$, there is at least one $Z \in \mathcal{A}(\Xi)$ with $\#Z = m(\Xi)+1$. The corresponding D_Z is of order $m(\Xi)+1$, hence cannot vanish identically on $\Pi_{m(\Xi)+1}$.

As for the sharpness of the second inclusion, recall from (I.28)Proposition that some polynomial pieces in M_Ξ are of exact degree $k(\Xi)$, while every such polynomial must be in $D(\Xi)$, by (I.37)Proposition. □

For example, the box-spline M_Ξ for $\Xi = [\mathrm{i}_1, \mathrm{i}_1, \mathrm{i}_2, \mathrm{i}_2] \in \mathbb{Z}^{2\times 4}$ is bilinear, and $k(\Xi) = 2$ and $m(\Xi) = 1$, since $\mathcal{A}_{\min}(\Xi) = \{[\mathrm{i}_1, \mathrm{i}_1], [\mathrm{i}_2, \mathrm{i}_2]\}$ in this case. Further, $D(\Xi) = \Pi_{1,1}$, and both inclusions in (60) are strict; see also (42)Tensor Products.

(62)Corollary. *The polynomials which make up M_Ξ span $D(\Xi)$.*

Proof. Any polynomial in $D(\Xi) = \Pi_M$ is necessarily a linear combination of the polynomials which make up $M = M_\Xi$, while, by (I.37)Proposition, each polynomial from M_Ξ lies in $D(\Xi)$. □

A basis for $D(\Xi)$. Here is a way to generate $\#\mathcal{B}(\Xi)$ linearly independent elements of $D(\Xi)$. Since $\dim D(\Xi) = \#\mathcal{B}(\Xi)$ by (32)Theorem, these elements necessarily form a basis for $D(\Xi)$. In fact, the simple argument for showing that $\dim D(\Xi) \le \#\mathcal{B}(\Xi)$ (see (39)), together with the present material, provides an alternative proof of (32)Theorem.

For $c \in \mathbb{C}^\Xi$ and $Z \subseteq \Xi$, define

$$D_{Z,c} := \prod_{\zeta \in Z} (D_\zeta - c(\zeta)),$$

and note that

$$D_{Z,c} e^{\vartheta \cdot} = e^{\vartheta \cdot} \prod_{\zeta \in Z} \left(\vartheta\zeta - c(\zeta)\right).$$

In particular, if $\vartheta\zeta = c(\zeta)$ for all ζ in some *basis* Z, then, for every $H \in \mathbb{H}(\Xi)$, some $\zeta \in \Xi \backslash H$ is necessarily in Z, and therefore $D_{\Xi\backslash H,c} e^{\vartheta \cdot} = 0$.

This shows that each $c \in \mathbb{C}^\Xi$ generates a map $\mathcal{B}(\Xi) \to \mathbb{C}^s : Z \mapsto \vartheta_Z$, from the *multiset* $\mathcal{B}(\Xi)$ into $\mathbb{C}^s$, with ϑ_Z the unique solution of the linear system $\zeta? = c(\zeta), \zeta \in Z$, and that the resulting exponentials $e^{\vartheta_Z \cdot}$ are all in $\cap_{H \in \mathbb{H}(\Xi)} \ker D_{\Xi\backslash H,c}$.

Note that ϑ_Z is the negative of the point x_Z used in the second proof, by way of zonotopes, of (I.53)Theorem. We claim again that, generically, $\vartheta_Z \ne \vartheta_{Z'}$ whenever $Z \ne Z'$ (even if the two are equal as matrices). Indeed, for any $\xi \in \Xi$, the set $\{\vartheta_Z : Z \in \mathcal{B}(\Xi)\backslash\mathcal{B}(\Xi\backslash\xi)\}$ must lie on the (shifted) hyperplane $\{\vartheta : \xi\vartheta = c(\xi)\}$, hence will contain any of the finitely many ϑ_Z with $Z \in \mathcal{B}(\Xi\backslash\xi)$ only for finitely many choices of $c(\xi)$.

We may therefore assume (what is true for almost any choice of c, viz.) that $\#\{\vartheta_Z : Z \in \mathcal{B}(\Xi)\} = \#\mathcal{B}(\Xi)$. Correspondingly,

$$F := \operatorname{span}\{e^{\vartheta_Z \cdot} : Z \in \mathcal{B}(\Xi)\}$$

has dimension $\#\mathcal{B}(\Xi)$, and

(63) $$F \subseteq \cap_{H \in \mathbb{H}(\Xi)} \ker D_{\Xi\backslash H,c}.$$

We now relate F to $D(\Xi)$ by the following observation. For any $f \in F$, consider its power series

$$f = f_0 + f_1 + f_2 + \dots$$

with the terms grouped by order, i.e., with f_j homogeneous of degree j, all j. In particular, single out the **initial term**, i.e., the first nonzero term in that grouped power series, and denote it by $f_\downarrow$ (in words: f 'least'). If $D_{Z,c} f = 0$, then necessarily already $D_Z(f_\downarrow) = 0$, since D_Z is the leading term, i.e., the term of highest order, in the differential operator $D_{Z,c}$. This implies with (63) that

$$F_\downarrow := \operatorname{span}\{f_\downarrow : f \in F\} \subseteq D(\Xi).$$

From this observation, we obtain a basis for $D(\Xi)$ as follows. Let

$$\Theta := \{\vartheta_Z : Z \in \mathcal{B}(\Xi)\}.$$

Since the exponentials $e^{\vartheta\cdot}$, $\vartheta \in \Theta$, are linearly independent, so must be the vectors $\big((\vartheta^\alpha)_{\alpha\in\mathbb{Z}_+^s}\big)$ of their normalized Taylor coefficients at the origin (note that $(D^\alpha e^{\vartheta\cdot})(0) = \vartheta^\alpha$). In other words, the matrix

$$V := \Big(\vartheta^\alpha\Big)_{\vartheta\in\Theta,\alpha\in\mathbb{Z}_+^s},$$

with rows indexed by $\vartheta \in \Theta$ and columns indexed by $\alpha \in \mathbb{Z}_+^s$, is of full rank. Thus elimination with partial pivoting implies that, for some ordering $\vartheta_1, \vartheta_2, \ldots$ of the elements of Θ and for any ordering of the elements of $\mathbb{Z}_+^s$, there is a factorization $V = LU$ of V with L unit lower triangular, and with U in row echelon form in the sense that there exists a strictly increasing sequence $\alpha_1 < \alpha_2 < \ldots$ (in the ordering we used in writing down V) so that, for all i, $U(\vartheta_i, \alpha_i)$ is the first nonzero entry in row $U(\vartheta_i, :)$. For reasons that will be obvious in a moment, we choose any ordering of $\mathbb{Z}_+^s$ consistent with degree, i.e., any ordering for which $\alpha \le \beta$ implies that $|\alpha| \le |\beta|$. Thus, the first column of V contains the zeroth order terms, the next s columns contain the s first-order terms, the next $\binom{s}{2}$ columns contain the second-order terms, etc.

Now, since $e^{\vartheta_i\cdot} = \sum_\alpha [\![]\!]^\alpha V(\vartheta_i, \alpha)$ and L is invertible, it follows that each

$$f_i := \sum_\alpha [\![]\!]^\alpha U(\vartheta_i, \alpha)$$

is also in F, and that the resulting sequence $f_1, f_2, \ldots$ is linearly independent since the power series for f_i has its first nonzero term later than any of the f_j with $j < i$. Moreover, the very same statement holds for the $f_{i\downarrow}$, hence the resulting sequence $f_{1\downarrow}, f_{2\downarrow}, \ldots$ must be linearly independent. Since it consists of $\#\mathcal{B}(\Xi)$ terms (by our choice of c) and lies in $D(\Xi)$, it must therefore be a basis for $D(\Xi)$ since $\dim D(\Xi) \le \#\mathcal{B}(\Xi)$ by (39). In fact, it follows from the construction process outlined that this basis is special in several ways: (i) it consists of homogeneous polynomials; (ii) for each j, it contains as many elements of degree $\le j$ as are possible in any basis for $D(\Xi)$; in particular, (iii) it contains the lowest possible number of elements of exact degree $k = k(\Xi)$.

There is no difficulty in constructing a basis for $D(\Xi)$ in the generic case, since then $D(\Xi) = \Pi_k$, by (31)Proposition. We therefore choose two simple examples in which $D(\Xi)$ fails to equal Π_k.

(64)Example. Consider $\Xi = [\mathbf{i}_1, \mathbf{i}_1, \mathbf{i}_2, \mathbf{i}_2] =: [\xi_1, \ldots, \xi_4]$. Then

$$\mathcal{B}(\Xi) = \{[\xi_1, \xi_3], [\xi_1, \xi_4], [\xi_2, \xi_3], [\xi_2, \xi_4]\}.$$

Consequently, $\vartheta_{[\xi_i,\xi_j]} = (c(i), c(j))$, for $i \le 2 < j$. For example, with $c = (0,1,0,1)$, we get

$$\Theta = \{(0,0),(0,1),(1,0),(1,1)\}.$$

The matrix V in this ordering of the rows and with the lexicographic ordering for $\mathbb{Z}_+^2$ takes the form

α	0,0	1,0	0,1	2,0	1,1	0,2	$\cdots$
(0,0)	1	0	0	0	0	0	$\cdots$
(0,1)	1	0	1	0	0	1	$\cdots$
(1,0)	1	1	0	1	0	0	$\cdots$
(1,1)	1	1	1	1	1	1	$\cdots$

Elimination with partial (row) pivoting applied to this matrix interchanges rows 2 and 3, and so supplies the factorization

$$\begin{bmatrix} 1 & 0 & 0 & 0 & 0 & 0 & \cdots \\ 1 & 1 & 0 & 1 & 0 & 0 & \cdots \\ 1 & 0 & 1 & 0 & 0 & 1 & \cdots \\ 1 & 1 & 1 & 1 & 1 & 1 & \cdots \end{bmatrix} = V = LU =$$

$$= \begin{bmatrix} 1 & 0 & 0 & 0 \\ 1 & 1 & 0 & 0 \\ 1 & 0 & 1 & 0 \\ 1 & 1 & 1 & 1 \end{bmatrix} \begin{bmatrix} 1 & 0 & 0 & 0 & 0 & 0 & \cdots \\ 0 & 1 & 0 & 1 & 0 & 0 & \cdots \\ 0 & 0 & 1 & 0 & 0 & 1 & \cdots \\ 0 & 0 & 0 & 0 & 1 & 0 & \cdots \end{bmatrix}$$

From this, we read off the basis $f_{1\downarrow} = [\![]\!]^0$, $f_{2\downarrow} = [\![]\!]^{1,0}$, $f_{3\downarrow} = [\![]\!]^{0,1}$, $f_{4\downarrow} = [\![]\!]^{1,1}$ for $D(\Xi) = \Pi_{1,1}$.

Now consider $\Xi = [\mathrm{i}_1 + \mathrm{i}_2, \mathrm{i}_2]_{(1,3)} = [\mathrm{i}_1 + \mathrm{i}_2, \mathrm{i}_2, \mathrm{i}_2, \mathrm{i}_2] =: [\xi_1, \ldots, \xi_4]$, hence $\mathcal{B}(\Xi) = \{[\xi_1,\xi_2],[\xi_1,\xi_3],[\xi_1,\xi_4]\}$, and $\vartheta_{[\xi_1,\xi_i]} = ([\mathrm{i}_1 + \mathrm{i}_2, \mathrm{i}_2]^T)^{-1}(c(1), c(i)) = (c(1) - c(i), c(i))$. Thus, with $c = (0,0,1,2)$, we get the three points $\Theta = \{(0,0),(-1,1),(-2,2)\}$.

For this example, V takes the form

α	0,0	1,0	0,1	2,0	1,1	0,2	$\cdots$
(0,0)	1	0	0	0	0	0	$\cdots$
(−1,1)	1	−1	1	1	−1	1	$\cdots$
(−2,2)	1	−2	2	4	−4	4	$\cdots$

Elimination without pivoting can be applied to this matrix and supplies the factorization

$$\begin{bmatrix} 1 & 0 & 0 & 0 & 0 & 0 & \cdots \\ 1 & -1 & 1 & 1 & -1 & 1 & \cdots \\ 1 & -2 & 2 & 4 & -4 & 4 & \cdots \end{bmatrix} = V =$$

$$= \begin{bmatrix} 1 & 0 & 0 \\ 1 & 1 & 0 \\ 1 & 2 & 1 \end{bmatrix} \begin{bmatrix} 1 & 0 & 0 & 0 & 0 & 0 & \cdots \\ 0 & -1 & 1 & 1 & -1 & 1 & \cdots \\ 0 & 0 & 0 & 2 & -2 & 2 & \cdots \end{bmatrix}.$$

From this, we read off the basis $f_{1\downarrow} = [\![]\!]^0$, $f_{2\downarrow} = \zeta\cdot$, $f_{3\downarrow} = (\zeta\cdot)^2$ for $D(\Xi)$, with $\zeta := (-1, 1)$. In other words, for this example, $D(\Xi)$ is $\Pi_{0,2}$ rotated 45°, hence constant in the direction $\mathbf{i}_1 + \mathbf{i}_2$. This could have been deduced directly from the fact that, in this example, $\mathcal{A}(\Xi) = \{[\mathbf{i}_2, \mathbf{i}_2, \mathbf{i}_2], [\mathbf{i}_1 + \mathbf{i}_2]\}$, hence $D(\Xi) = \ker(D_2)^3 \cap \ker(D_1 + D_2)$. ▣

(65)Remark: *D*- and *σ*-invariance. We briefly address the claim that (43)Figure gives a qualitatively correct description of certain important features of $D(\Xi)$ for arbitrary Ξ, viz. the features due to the differentiation- and scale-invariance of $D(\Xi)$. Since $D(\Xi)$ is the joint kernel of constant-coefficient differential operators, it is D-invariant, i.e., it contains all derivatives of p if it contains p. Further, since these differential operators are homogeneous, $D(\Xi)$ is σ-invariant, i.e., it contains $p(\cdot/h)$ for any h if it contains p. Equivalently, $D(\Xi)$ is spanned by homogeneous polynomials.

Since $D(\Xi)$ is finite-dimensional, it must contain **maximal elements**, i.e., homogeneous polynomials which themselves are not proper linear combinations of derivatives of polynomials in $D(\Xi)$. Further, it must contain a **complete maximal set**, i.e., a maximally linearly independent set of maximal elements. Any such complete maximal set is a **minimal generating set for** $D(\Xi)$ in the sense that $D(\Xi)$ is the sum of the polynomial spaces obtained as the span of all the derivatives (including the zeroth order one) of one of those maximal elements. Further, any two complete maximal sets contain the same number of elements (since this is true for any two maximally linearly independent sets in some subset of a finite-dimensional linear space). Equivalently, a set of homogeneous polynomials in $D(\Xi)$ is a minimal generating set for $D(\Xi)$ exactly when it is a basis for some algebraic complement in $D(\Xi)$ of the subspace $D(\Xi)' := \sum_{|\alpha|=1} D^\alpha D(\Xi)$.

Finally, any maximal element of $D(\Xi)$ has degree $k = k(\Xi)$. Indeed, $D(\Xi)$ is spanned by the polynomials which provide the polynomial pieces from which M_Ξ is made, and, by (I.29), $M_\Xi = \nabla_\Xi T_\Xi$, hence the polynomial pieces which form T_Ξ and their shifts or, equivalently, their derivatives, span $D(\Xi)$. Since the polynomial pieces which make up T_Ξ are all of exact degree k, it follows that every maximal element of $D(\Xi)$ must be of that degree. In terms of (43)Figure, this means that all 'peaks' of $D(\Xi)$ have their apex on the same level.

Notes. Linear combinations of shifts of box splines appear in the first paper on box splines, [de Boor, DeVore'83]. However, the tradition of studying approximation from a **principal shift-invariant space**, i.e., a space generated by the shifts of one function, goes back at least to [Schoenberg'46a] whose resulting beautiful theory of *univariate* cardinal splines is summarized and detailed in [Schoenberg'73a]. It also plays a prominent role in the analysis of the finite element method; see, e.g., [Fix, Strang'69], [Strang'71], [Strang, Fix'73a-b], and references therein. The term 'principal shift-invariant space' comes from [de Boor, DeVore, Ron'92a].

As to the notational conventions used here for the various *convolutions*, we have stuck with the meaning, introduced in [de Boor'87c], of $M*'f$ as the semi-discrete convolution of the two *functions* M and f, in order to distinguish it from the convolution $M*f$ in which one integrates or sums, as the case may be, over the set common to the domains of M and f. In particular, contrary to recent usage, we do not use the notation $M*'a$ for the (semi-discrete) convolution of M with the *sequence* a.

The results in the section on the *partition of unity* are taken from [de Boor, Höllig'82b]. The 'standard argument' alluded to in that section is the following: One observes that, for a compactly supported nonnegative M with $\sum_{j\in\mathbb{Z}^s} M(\cdot - j) = 1$, for any x, $|f(x) - \sum_{j\in\mathbb{Z}^s} M(x-j)f(j)| = |\sum_{j\in\mathbb{Z}^s} M(x-j)(f(x)-f(j))| = |\sum_{M(x-j)\neq 0} M(x-j)(f(x)-f(j))| \le \max_{M(x-j)\neq 0}|f(x)-f(j)| \le \omega_f(\operatorname{diam}\operatorname{supp} M)$, with ω_f the modulus of continuity of f. Hence, for a uniformly continuous f, $\sup_{x\in\mathbb{R}^s}|f(x) - \sum_{j\in\mathbb{Z}^s} M(x/h-j)f(hj)| \le \omega_f(h\text{const}) \to 0$ as $h\to 0$, since const $:= \operatorname{diam}\operatorname{supp} M < \infty$, by assumption.

The *box spline shifts which do not vanish at a given point* were first given attention in [Dahmen, Micchelli'85d]. In particular, (15)Lemma is from there. A set like $\iota(x,\Xi)$ was introduced there and denoted $b(x,\Xi)$. To be precise, $b(x,\Xi)$ is defined there as $\{j\in\mathbb{Z}^s : M(x-j)\neq 0\}$, which, by (2), is, in general, only a subset of $\iota(x,\Xi)$. While the two sets do coincide whenever $x\notin\Gamma(\Xi)$, the set $\iota(x,\Xi)$ was found to be slightly more convenient in the arguments in Chapter VI. In fact, it was more convenient even than the 'natural' definition $(x-\Xi\square)\cap\mathbb{Z}^s$ which would have made unnecessary the restriction of x to the complement of $\Gamma(\Xi)$ as enforced throughout this chapter whenever $\iota(x,\Xi)$ is mentioned. We switched, from $b(x,\Xi)$ to $\iota(x,\Xi)$, since $b(x|\Xi)$ is used in the literature for the value at x of the *discrete* box spline with directions Ξ.

The *differentiation* formula (17) is from [de Boor, Höllig'82b] as is the fact that unimodularity of Ξ is a necessary condition for the *linear independence* of the shifts of a box spline. The sufficiency of this condition was proved in [Jia'84b] and, independently, in [Dahmen, Micchelli'83b]. The fact that, for box splines, (global) linear independence implies local linear independence was proved in [Jia'85] and [Dahmen, Micchelli'85d]. This implication does not hold in general (cf. [BenArtzi, Ron'90]).

Dahmen and Micchelli were the first to consider the characterization of the dependence relations for the shifts of a box spline. A version of their resulting beautiful and remarkable theory (announced in [Dahmen, Micchelli'84d] and detailed in [Dahmen, Micchelli'85c,'85d]), is given here in (32)Theorem, (46)Theorem, (49)Theorem, (55)Theorem, and (57)Theorem (, not necessarily with the original proofs). This is the *centerpiece* of box spline theory. Without such a theory in place, this book would very likely not have been written.

Note that the theory presented here does not, in general, describe the space of *all* dependence relations for the shifts of a box spline. In other words, $\ker M^\star$ is usually much larger than $E(\Xi)$. For example, if $M = M_1 * M_2$, then both $\ker M_i$ are contained in $\ker M^\star$. Also, if $M = M_1 \otimes M_2$, then $\ker M^\star$ contains $\ker M_1 \otimes$ *anything*. Also, in general, $\ker M^\star$ may contain exponentials without containing bounded exponentials. This makes (57)Theorem(vi) all the more remarkable since it states that, for M a box-spline, $\ker M^\star$ is trivial if and only if it contains no bounded exponential.

An alternative proof of (32)Theorem, in the following, more precise, form, has been given in [Hakopian, Saakyan'88].

(66)Theorem. *Let Ξ be a matrix with columns from $\mathbb{R}^s\backslash 0$, and consider the problem of finding, for given smooth functions ψ and φ_H, all $H \in \mathbb{H}(\Xi)$, a function f which satisfies*

$$\text{(67)} \qquad \begin{aligned} &\forall\{H \in \mathbb{H}(\Xi)\} \; D_{\Xi\backslash H} f = \varphi_H \\ &\forall\{Z \subseteq \Xi : \operatorname{ran} \Xi\backslash Z = \mathbb{R}^s\} \; D_Z f(0) = D_Z \psi(0). \end{aligned}$$

If the φ_H satisfy the obviously necessary compatibility condition

$$\text{(68)} \qquad \forall\{H \neq H'\} \; D_{(\Xi\cap H)\backslash H'} \varphi_H = D_{(\Xi\cap H')\backslash H} \varphi_{H'},$$

then (67) has exactly one solution.

In particular, with the obviously compatible choice $\varphi_H = 0$, all $H \in \mathbb{H}(\Xi)$, this theorem claims that $D(\Xi) = \cap_{H\in\mathbb{H}(\Xi)} \ker D_{\Xi\backslash H}$ has dimension equal to that of the polynomial space

$$P(\Xi) := \operatorname{span}\{p_Z : Z \subseteq \Xi, \operatorname{ran} \Xi\backslash Z = \mathbb{R}^s\}.$$

Here, $p_Z := \prod_{\zeta\in Z}(\zeta\,\cdot\,)$ is the homogeneous polynomial obtained as the product of the linear factors $x \mapsto \zeta x$, $\zeta \in Z$. The space $P(\Xi)$ is readily seen to have dimension equal to $\#\mathcal{B}(\Xi)$, thus (32)Theorem follows. Quite independently, [Jia'90e] also proves that $P(\Xi)$ is dual to $D(\Xi)$, as do [Dyn, Ron'90].

The fact that $D(\Xi)$ coincides with *the polynomials in* S can already be found in [de Boor, Höllig'82b], as can the statement (60). We added the rest of (59)Proposition, for completeness. We cite from [Ron, Sivakumar'93]the following extension of (59)Proposition, in order to illustrate the complications that can arise when one permits the directions in Ξ to be *rational.*

(69)Proposition. *If* Ξ *is a* rational *matrix with* $\operatorname{ran}\Xi = \mathbb{R}^s$, *then*

$$\Pi_M = \bigcap_{Z \in \mathcal{A}_{\mathrm{rat}}(\Xi)} \ker D_Z,$$

with $\mathcal{A}_{\mathrm{rat}}(\Xi)$ *consisting of those* $Z \subseteq \Xi$ *for which the shifts of* $M_{\Xi\backslash Z}$ *fail to form a partition of unity.*

While this proposition gives a formula for Π_M (of use for the determination of approximation order as detailed in the next chapter), there is at present no formula for the dimension of Π_M for this case.

The idea that $D(\Xi)$ can be understood much more simply by perturbing the differential operators involved is a major contribution of *exponential* box spline theory to box spline theory. It occurs first in [BenArtzi, Ron'88]. The description of $D(\Xi)$ as the 'least' of some exponential space is taken from [de Boor, Ron'91a], while the particular construction of a basis for such a 'least' comes from [de Boor, Ron'92b]. Since the polynomials which make up M_Ξ provide a spanning set for $D(\Xi)$ (by (62)Corollary), and can be written as linear combinations of shifts of the truncated power T_Ξ (by (I.29)), it is also possible to generate a basis for $D(\Xi)$ from the shifts of T_Ξ; see [Dahmen'90].

III

Quasi-interpolants & approximation power

In this chapter, we consider the **approximation power** of the cardinal spline space

$$S = S_M = \operatorname{span}\Big(M(\cdot - j)\Big)_{j\in\mathbb{Z}^s} = \operatorname{ran} M*'$$

defined in (II.1). We measure this approximation power in the following way: Consider the **scale** (S_h) associated with S, where

$$S_h := \sigma_h(S),$$

with

$$\sigma_h f : x \mapsto f(x/h).$$

Thus, S_h is a cardinal spline space on the refined lattice $h\mathbb{Z}^s$. We define the **approximation order** of S to be the largest r for which

$$\operatorname{dist}(f, S_h) = O(h^r)$$

for all sufficiently smooth f, with the distance measured in the $\mathbf{L}_p(G)$-norm $(1 \le p \le \infty)$ on some bounded domain G with piecewise smooth boundary.

We show that, for any such choice of p and G, the approximation order is $m(\Xi) + 1$, with

$$m(\Xi) + 1 = \min\{\#\mathrm{Z} : \mathrm{Z} \in \mathcal{A}(\Xi)\}$$

the largest integer r for which $\Pi_{<r} \subseteq D(\Xi)$, according to (II.59)Proposition. We also give a general recipe for approximation schemes (called quasi-interpolants) which realize this approximation order.

An upper bound. The approximation order of the spline space S cannot be any better than the approximation order at 0 of the polynomial space $D(\Xi)$.

(1)Proposition. *The approximation order of S_M does not exceed $m(\Xi)+1$.*

Proof. Let U be an open set in $\mathbb{R}^s$. Then for any $p \in [1\,..\,\infty]$ and any $Z \subseteq \Xi$,

$$\|D_Z f\|_{\mathbf{L}_p(U)} \le c\|f\|_{\mathbf{L}_p(U)}, \qquad \forall f \in \Pi_m,$$

for some c, since D_Z is linear and Π_m is finite-dimensional. Under the change σ_h in scale, this inequality becomes

$$(2) \qquad \|D_Z f\|_{\mathbf{L}_p(hU)} \le \frac{c}{h^{\#Z}}\|f\|_{\mathbf{L}_p(hU)}, \qquad \forall f \in \Pi_m.$$

This formula will be applied to each of the connected components, U, of $\mathbb{R}^s\backslash\Gamma(\Xi)$. Up to translations by $\mathbb{Z}^s$, there are only finitely many of these regions. Since Π_m is **translation-invariant** (i.e., contains $p(\cdot + x)$ for any $p \in \Pi_m$ and any $x \in \mathbb{R}^s$), we may therefore assume that (2) holds with the same c on each of these sets. Also, let

$$G_h := \bigcup_{hU \subset G} hU,$$

where U runs over all connected components of $\mathbb{R}^s\backslash\Gamma(\Xi)$.

By definition of $m(\Xi)$, there is a $Z \in \mathcal{A}(\Xi)$ with $\#Z = m(\Xi)+1$ and a polynomial $f \in \Pi_{m(\Xi)+1}$ with $D_Z f = 1$. Choose $Z' \subseteq Z$ with $\#Z' = m(\Xi)$ and set $z := Z\backslash Z'$. If, for some $p \in [1\,..\,\infty]$ and some approximating sequence $g_h \in S_h$, we have

$$\|g_h - f\|_{\mathbf{L}_p(G)} = o(h^{m(\Xi)+1}),$$

then also $\|g_h - f\|_{\mathbf{L}_1(G)} = o(h^{m(\Xi)+1})$, and it follows from (2) that

$$\begin{aligned} \|D_{Z'}g_h - D_{Z'}f\|_{\mathbf{L}_1(G_h)} &= \sum_{hU\subset G} \|D_{Z'}g_h - D_{Z'}f\|_{\mathbf{L}_1(hU)} \\ (3) \qquad &\le \left(c/h^{m(\Xi)}\right) \sum_{hU\subset G} \|g_h - f\|_{\mathbf{L}_1(hU)} \\ &\le \left(c/h^{m(\Xi)}\right) \|g_h - f\|_{\mathbf{L}_1(G)} \;= o(h). \end{aligned}$$

Since the support of $D_z D_{Z'} g_h$ is contained in hyperplanes (viz., in $\Gamma(\Xi)$), $D_{Z'} g_h$ is piecewise constant on lines in the direction z. On the other hand, $D_{Z'} f$ restricted to these lines is of the form $(D_{Z'} f)(x+tz) = (D_{Z'} f)(x) + t$, i.e., it is linear in the direction of z. Each U contains in its interior a cube V with one side parallel to z. Since the error in $\mathbf{L}_1([a\,..\,b])$ in approximating a linear function of slope 1 by constants is $\ge \text{const}\,(b-a)^2$, $\|D_{Z'} g_h - D_{Z'} f\|_{\mathbf{L}_1(hV)} \ge \text{const}\, h^{s-1} h^2$. Since $hV \subset hU$ and the number of hU in G is of order $1/h^s$, we find that

$$\|D_{Z'} g_h - D_{Z'} f\|_{\mathbf{L}_1(G_h)} \ge \text{const}\, h,$$

in contradiction to (3). ☐

This upper limit on the approximation order is actually attained by a local quasi-interpolant, as we show next, in a sequence of short sections. In this discussion, the only properties of the box spline M used are that M is a compactly supported piecewise continuous function.

Quasi-interpolants. A **quasi-interpolant** Q for S is a linear map into S which is local, bounded (in some relevant norm), and reproduces some (nontrivial) polynomial space. We say that Q has **polynomial order** r, in case Q **reproduces** $\Pi_{<r}$, i.e., in case $Q = 1$ on $\Pi_{<r}$. We restrict attention to quasi-interpolants of the form

$$Q_\lambda f(x) := \sum_j M(x-j)\lambda f(\cdot + j)$$

in which λ is some suitable linear functional. For example, if $\lambda = \delta_0$, then $Q_\lambda = M*'$, and this is of polynomial order ≥ 1 for any box spline M, since the integer translates of any box spline form a partition of unity, by (II.10)Theorem.

Here is the basic observation regarding quasi-interpolants.

(4)Proposition. *Let B_R be a ball of radius R centered at the origin and let G be a domain in $\mathbb{R}^s$. If λ is a continuous linear functional on $\mathbf{L}_p(B_R)$ for some positive R and some $p \in [1\,..\,\infty]$, and $Q = Q_\lambda$ is of polynomial order r, then*

$$\|\sigma_h Q \sigma_{1/h} f - f\|_{\mathbf{L}_p(G)} = O(h^r)$$

for any sufficiently smooth function f (e.g., for $f \in C^r(G + B_\varepsilon)$ for some $\varepsilon > 0$).

Proof. We consider the error on hU, with U any connected component of $\mathbb{R}^s \backslash \Gamma(\Xi)$.

Since $\sigma_h Q \sigma_{1/h} f = \sum_j M(\cdot/h - j)\lambda f(h(\cdot + j))$, it is, on hU, entirely determined by those j for which $M(\cdot/h - j)$ has some support in hU, i.e., for which $hj \in h(U - \Xi\square)$. These j make up the set

$$J_U := \iota(U, \Xi) = \cup_{y \in U} \iota(y, \Xi),$$

whose cardinality is independent of U since $J_U = \iota(y, \Xi)$ for any $y \in U$. For the coefficients $\lambda(f(h(\cdot + j)))$, $j \in J_U$, we get the estimate

$$\begin{aligned} |\lambda f(h(\cdot + j))| &\le \|\lambda\| \|f(h(\cdot + j))\|_{L_p(B_R)} \\ &\le \|\lambda\|\, h^{-s/p}\, \|f(\cdot + hj)\|_{L_p(hB_R)} \le \|\lambda\|\, h^{-s/p}\, \|f\|_{L_p(h(U+B_{R'}))}, \end{aligned}$$

with R' chosen so that $B_R - \Xi\square \subseteq B_{R'}$.

With this uniform estimate on the coefficients relevant to hU in hand, the fact that the box spline shifts $M(\cdot/h - j)$ form a local partition of unity provides a 'local' error estimate: Indeed,

$$\sum_{j \in J_U} M(\cdot/h - j)_{|hU} = 1,$$

and

$$\sigma_h Q \sigma_{1/h} f_{|hU} = \sum_{j \in J_U} M(\cdot/h - j)_{|hU} \lambda f(h(\cdot + j)).$$

Consequently,

$$|\sigma_h Q \sigma_{1/h} f| \le \|\lambda\|\, h^{-s/p}\, \|f\|_{L_p(h(U+B_{R'}))} \qquad \text{on} \quad hU.$$

Since Q is of polynomial order r, it follows that, for any $g \in \Pi_{<r}$,

$$\sigma_h Q \sigma_{1/h} f - f = (\sigma_h Q \sigma_{1/h} - 1)(f - g) \qquad \text{on} \quad hU.$$

Combining the last two observations with the fact $\text{meas}(hU) = h^s \,\text{meas}(U) \le \text{const}\, h^s$, we find

$$\|\sigma_h Q \sigma_{1/h} f - f\|_{L_p(hU)} \le (\text{const}\, \|\lambda\| + 1)\|f - g\|_{L_p(h(U+B_{R'}))}, \qquad \forall\, g \in \Pi_{<r}.$$

Finally, since $\text{dist}(f, \Pi_{<r})_{\mathbf{L}_p(V)} \le \text{const} \sum_{|\alpha|=r} \|D^\alpha f\|_{\mathbf{L}_p(V)} \big(\text{diam}(V)\big)^r$, we have

$$\|\sigma_h Q \sigma_{1/h} f - f\|_{\mathbf{L}_p(hU)} \le \text{const}\, h^r \sum_{|\alpha|=r} \|D^\alpha f\|_{\mathbf{L}_p(h(U+B_{R'}))}.$$

The global estimate now follows for $p < \infty$ by summing this local estimate over all hU with $hU \cap G \ne \emptyset$, thus getting

$$\|\sigma_h Q \sigma_{1/h} f - f\|_{\mathbf{L}_p(G)} \le \text{const}\, h^r \sum_{|\alpha|=r} \|D^\alpha f\|_{\mathbf{L}_p(G+B_{R'})} = O(h^r).$$

The same conclusion is reached for $p = \infty$ by maximizing the local estimate over all such hU instead. □

It follows that the approximation order of our cardinal spline space is (at least) r provided we can determine a linear functional λ which is bounded on some $\mathbf{L}_p(B_R)$ and such that Q_λ reproduces $\Pi_{<r}$.

For this, we note that we can write Q_λ equivalently as

$$Q_\lambda = M*'\nu*,$$

with $\nu := \lambda\sigma_{-1}$, i.e., with $\nu f := \lambda f(-\cdot)$. This has some advantages since, as we will see in a moment, the semi-discrete convolution $M*'$ can be written on S as an *ordinary* convolution, and this makes it easy to provide for it a right inverse on S. Precisely, we will show that $M*'$ agrees on S with the (ordinary) convolution $M^{|}*$, with $M^{|}$ our notation for the *linear functional*

$$(5) \qquad M^{|} := \sum_j M(j)\delta_j$$

and not to be confused with the mesh-function $M_| = M_{|\mathbb{Z}^s}$. Further, in contrast to $M*'$, $M^{|}*$ maps all of Π into Π. Since we would like to have $Q_\lambda = 1$ on $\Pi_{<r}$, we are, in effect, hoping to choose ν so that $\nu*$ is a (right) inverse for $M*'$ on $\Pi_{<r}$, and this we can achieve by making $\nu*$ a right inverse of $M^{|}*$ on some suitable polynomial space containing $\Pi_{<r}$. The following facts about convolutions and polynomials are therefore of interest.

Convolutions on polynomial spaces. The first proposition below makes clear that any convolution $\nu*$ is weakly degree-preserving in that $\nu*(\Pi_\alpha) \subseteq \Pi_\alpha$ for any α, while the second makes clear that a **translation-invariant** linear map (i.e., a map that commutes with any translation) can always be represented as a convolution.

(6)Proposition. *For any convolution $\nu*$ and any α, $\nu*(\Pi_\alpha) \subseteq \Pi_\alpha$, with equality if and only if $\nu*[\![\,]\!]^0 \neq 0$. In particular, $\nu*$ is invertible on Π (i.e., 1-1 and onto as a map from Π to Π) in case $\nu*[\![\,]\!]^0 \neq 0$.*

Proof. Since $\nu*f : x \mapsto \nu f(x-\cdot)$ and $[\![x-\cdot]\!]^\alpha = \sum_{\beta+\gamma=\alpha} [\![x]\!]^\beta [\![-\cdot]\!]^\gamma$, we have

$$\nu*[\![\,]\!]^\alpha = \sum_{\beta+\gamma=\alpha} [\![\,]\!]^\beta \nu[\![-\cdot]\!]^\gamma,$$

which shows that $\nu*(\Pi_\alpha) \subseteq \Pi_\alpha$ for any α. It shows, more explicitly, that

$$\nu*f \in \left(\nu*[\![\,]\!]^0\right) f + \Pi_{<\deg f}$$

for any $f \in \Pi$, hence $\nu*$ is **degree-preserving** in case $\widehat{\nu}(0) := \nu*[]^0 \neq 0$. In that case, $U := 1 - (\nu*)/\widehat{\nu}(0)$ is **degree-reducing**, hence $U^{|\alpha|+1} = 0$ on Π_α, therefore $1 = 1 - \big(1 - (\nu*)/\widehat{\nu}(0)\big)^{|\alpha|+1} = (\nu*)A$ (with $A := q(\nu*)$ a polynomial in $\nu*$), and thus $\nu*$ is invertible on Π_α, for any α. Since $[]^0 \in \Pi_\alpha\backslash 0$ for any α, having $\nu*[]^0 \neq 0$ is certainly a necessary condition for the invertibility of $\nu*$ on Π_α. ☐

Since $\nu[\![- \cdot]\!]^\gamma = [\![- iD]\!]^\gamma \widehat{\nu}(0)$ in terms of the Fourier transform $\widehat{\nu}$ of ν, we have the following corollary of use later.

(7)Corollary. *For any convolution $\nu*$ and any α,*

$$\nu*[]^\alpha = [\![\cdot -iD]\!]^\alpha \widehat{\nu}(0) = \sum_\beta []^\beta\, [\![- iD]\!]^{\alpha-\beta} \widehat{\nu}(0).$$

(8)Proposition. *Let $A : F \to F$ be a translation-invariant linear map on some linear space F of functions on $\mathbb{R}^s$ which is invariant under translations as well as under the map $\sigma_{-1} : f \mapsto f(-\cdot)$. Then $A = \nu*$, with $\nu := \delta_0 A\sigma_{-1}$.*

Proof. By definition of ν, we have

$$\nu*f(x){=}\nu f(x - \cdot){=}\delta_0 A\sigma_{-1} f(x - \cdot){=}\delta_0 A f(x + \cdot){=}\delta_0 (Af)(x + \cdot){=}(Af)(x)$$

for arbitrary $f \in F$ and arbitrary $x \in \mathbb{R}^s$, with the translation invariance of A justifying the second-last equality. ☐

These propositions imply the following concerning the problem of choosing ν so that $M*'\nu* = 1$ on $\Pi_{<r}$: Since $\nu*$ carries $\Pi_{<r}$ into itself by (6), $M*'$ must map $\Pi_{<r}$ into itself, hence onto itself, i.e., $M*'$ must be invertible on $\Pi_{<r}$. This implies that $\Pi_{<r} \subseteq \Pi_M$, hence (not surprisingly in view of the upper bound provided by (1)Proposition) we cannot hope to solve this problem unless $r \le m(\Xi) + 1$.

Further, $M*'$ must be translation-invariant on $\Pi_{<r}$ (since any $\nu*$ is). Therefore, finally, by (8), ν must agree on $\Pi_{<r}$ with the linear functional $\delta_0\,(M*'_{|\Pi_{<r}})^{-1}\sigma_{-1}$. In terms of $\lambda = \nu\sigma_{-1}$, this says that Q_λ has polynomial order r provided λ agrees on $\Pi_{<r}$ with the linear functional $\delta_0\,(M*'_{|\Pi_{<r}})^{-1}$.

This leaves us with the problem of constructing the linear functional $\delta_0\,(M*'_{|\Pi_{<r}})^{-1}$. For this, it is very convenient to know that, on Π_M, $M*'$ agrees with the map $f \mapsto f*'M$, i.e., with the convolution $(\sum_j M(j)\delta_j)*$, as we show next.

The semi-discrete convolution is a convolution on its range. While the semi-discrete convolution product $M*'f$ is asymmetric, this asymmetry is not all that strong since, after all,

$$(9) \qquad M*'f = f*'M \quad \text{on } \mathbb{Z}^s.$$

This implies that

$$(10) \qquad M{*}'f = f{*}'M \quad \text{for all } f \in S,$$

since, for $f = M{*}a$,

$$\begin{aligned} M{*}'f &= M{*}(M_{|}{*}a) \\ &= M{*}(a{*}M_{|}) \\ &= (M{*}a){*}M_{|} = f{*}'M. \end{aligned}$$

In other words, $M{*}'$ agrees on S with the *convolution* $M^{|}{*}$, where, to recall, $M^{|} = \sum_j M(j)\delta_j$.

If now $\sum_j M(j) = 1$ (as is the case for any box spline, by (II.10) Theorem), then $M^{|}{*}[]^0 = 1$, hence $M^{|}{*}$ is then 1-1 on any polynomial space. In particular, $M^{|}{*}$ is invertible on $\Pi_M = \Pi \cap S$, on which it agrees with $M{*}'$.

Hence, any formula for $(M^{|}{*})^{-1}$ on Π is certain to provide also a formula for $(M{*}')^{-1}$ valid on any polynomial subspace of Π_M, and that is really all we need for the construction of quasi-interpolants.

Since $M^{|}{*}$ is a convolution, we can compute $(M^{|}{*})^{-1}f$ for any polynomial f as soon as we know the polynomials $g_\alpha := (M^{|}{*})^{-1}[]^\alpha$ for all α. This leads us to Marsden's identity.

Marsden identity. We know from the proof of (II.59)Proposition that

$$L := M{*}'_{|\Pi_M}$$

maps Π_M 1-1 and onto itself. This implies that we can obtain the coefficients a in the expansion $M{*}a := p \in \Pi_M$ of a polynomial as a linear combination of shifts of M as $a = (L^{-1}p)_{|}$. Unfortunately, we know of no general formula for (the action of) L^{-1}. But it is possible to describe a simple algorithm for the generation of the polynomials $g_\alpha := L^{-1}[]^\alpha$. We follow custom and call the resulting formula

$$(11) \qquad []^\alpha = M{*}'g_\alpha \quad \text{for } []^\alpha \in \Pi_M$$

a 'Marsden identity', the name initially given to any formula which describes a monomial (or perhaps a shifted monomial) as a linear combination of B-splines.

By (10), (11) is equivalent to the equation

$$[]^\alpha = M^{|}{*}g_\alpha \quad \text{for } []^\alpha \in \Pi_M.$$

But, in contrast to $M*'$, the map $M^{|}*$ carries *all* of Π into Π, in fact, is invertible (as a map from Π to Π). This implies that there exists, for any $\alpha \in \mathbb{Z}_+^s$, a (unique) polynomial g_α so that

$$(12) \qquad [\![]\!]^\alpha = M^{|}*g_\alpha.$$

Further, assuming that we know these polynomials, we can compute

$$(M^{|}*)^{-1}f = (M^{|}*)^{-1}\Big(\sum_\alpha [\![]\!]^\alpha D^\alpha f(0)\Big) = \sum_\alpha g_\alpha D^\alpha f(0)$$

for any polynomial f. In particular, the linear functional $\delta_0(M*'_{|\Pi_M})^{-1}$ we seek coincides on Π_M with

$$(13) \qquad \lambda_0 := \delta_0(M^{|}*_{|\Pi})^{-1} = \sum_\alpha g_\alpha(0)\delta_0 D^\alpha : f \mapsto \sum_\alpha g_\alpha(0)(D^\alpha f)(0).$$

Quasi-interpolant summary. We now summarize the results of the preceding sections as they apply to the construction of quasi-interpolants satisfying the hypotheses (hence the conclusions) of (4)Proposition.

(14)Theorem. *If λ is any linear functional, bounded on $\mathbf{L}_p(B_R)$ for some positive R and agreeing with λ_0 of (13) on $\Pi_{<r}$ for some $r \le m(\Xi)+1$, then the corresponding quasi-interpolant Q_λ provides the approximation order r from S. Since this imposes only finitely many conditions on the linear functional λ, this implies that S has approximation order $m(\Xi)+1$.*

Practical application of this theorem requires that we have the polynomials g_α, or at least the numbers $g_\alpha(0)$, in hand. As it turns out (see (16) below), g_α is entirely determined by the sequence $(g_\beta(0))_{\beta\le\alpha}$.

Appell sequence. We now consider the construction of the polynomials g_α satisfying (12), under the assumption (valid for any box spline M) that $\sum_j M(j) = 1$.

This assumption implies that

$$g_0 = [\![]\!]^0.$$

Further, on differentiating (12), we find that

$$(15) \quad [\![]\!]^{\alpha-\beta} = D^\beta(M^{|}*g_\alpha) = D^\beta(g_\alpha*'M) = (D^\beta g_\alpha)*'M = M^{|}*(D^\beta g_\alpha),$$

hence we know now that $g_\alpha \in [\![]\!]^\alpha + \Pi_{<\alpha}$ and that

$$D^\beta g_\alpha = g_{\alpha-\beta}.$$

This implies that

(16) $$g_\alpha = \sum_{\beta \le \alpha} c(\alpha - \beta) []^\beta$$

for the mesh function

$$c : \alpha \mapsto g_\alpha(0),$$

which already appeared in (13).

For its numerical determination, we conclude from (15) that

(17) $$\mu D^\beta g_\alpha = \delta_\beta(\alpha),$$

with

$$\mu := \delta_0 M^| * = M^| \sigma_{-1} : f \mapsto \sum_j M(j) f(-j) = \sum_j M(-j) f(j).$$

This means that the g_α form the **Appell sequence** for the linear functional μ.

On substituting (16) into (17), we obtain

(18) $$\mu D^\beta \Big(\sum_{\gamma \le \alpha} []^\gamma c(\alpha - \gamma) \Big) = \delta_\beta(\alpha).$$

This is a linear system with a unit upper-triangular coefficient matrix, hence is uniquely solvable. Backsubstitution provides the formula

(19) $$g_\alpha = []^\alpha - \sum_{\beta \ne \alpha} \Big(\mu []^{\alpha - \beta} \Big) g_\beta.$$

This formula can also be verified directly by induction on α: For $\gamma < \alpha$,

$$\begin{aligned} \mu D^\gamma g_\alpha &= \mu D^\gamma []^\alpha - \textstyle\sum_{\beta \ne \alpha} \Big(\mu []^{\alpha-\beta} \Big) \mu D^\gamma g_\beta \\ &= \mu []^{\alpha - \gamma} - \mu []^{\alpha-\gamma} = 0 \end{aligned}$$

with the second equality by induction hypothesis, while $\mu D^\alpha g_\alpha = \mu D^\alpha []^\alpha = \mu []^0 = 1$. Note that we have, in effect, proved that (17) *defines* the g_α, hence the g_α must reflect any symmetries that μ might have.

There are at least two obvious ways to construct the g_α, or, equivalently, the mesh function $c = \big(g_\alpha(0) \big)$. One possibility is to use (19) to build up g_α by induction on the degree α, starting with $g_0 = 1$.

(20) Example: the ZP element. We illustrate the inductive construction of the Appell sequence by considering the ZP element, i.e., $M =$

M_Ξ with $\Xi = \left[\begin{smallmatrix}1 & 0 & 1 & -1\\ 0 & 1 & 1 & 1\end{smallmatrix}\right]$. By symmetry, the values of M at the mesh points are given by

$$M(j) = \begin{cases} 1/4, & \text{for } j = (0,1), (1,1), (0,2), (1,2); \\ 0, & \text{otherwise.} \end{cases}$$

Therefore, $\mu = \delta_0 M^{|} *$ is in this case given by the rule

$$\mu f = (f(0,-1) + f(-1,-1) + f(0,-2) + f(-1,-2))/4.$$

With this, the algorithm based on (19) gives, in order,

$$\begin{aligned} g_{0,0} &= 1; \\ g_{1,0} &= [\![\,]\!]^{1,0} - \mu[\![\,]\!]^{1,0} 1 = [\![\,]\!]^{1,0} + 1/2; \\ g_{0,1} &= [\![\,]\!]^{0,1} - \mu[\![\,]\!]^{0,1} 1 = [\![\,]\!]^{0,1} + 3/2; \\ g_{1,1} &= [\![\,]\!]^{1,1} - \mu[\![\,]\!]^{1,1} 1 - \mu[\![\,]\!]^{1,0} g_{0,1} - \mu[\![\,]\!]^{0,1} g_{1,0} \\ &= [\![\,]\!]^{1,1} - (3/4)1 + (1/2) g_{0,1} + (3/2) g_{1,0} \\ &= g_{1,0} g_{0,1}, \end{aligned}$$

and

$$g_{2,0} = [\![\,]\!]^{2,0} + [\![\,]\!]^{1,0}/2, \quad g_{0,2} = [\![\,]\!]^{0,2} + (3/2)[\![\,]\!]^{0,1} + 1.$$

□

Another approach to the construction of the first few terms in the Appell sequence (g_α) is to solve (18) for one sufficiently large α, thereby obtaining simultaneously sufficient information about all g_γ with $\gamma \le \alpha$, because of (16). This requires generation of the matrix entries $\mu D^\beta [\![\,]\!]^\gamma = \mu [\![\,]\!]^{\gamma-\beta}$ (and a sensible ordering of the multi-indices involved). If, for our specific μ, the numbers $M(-j)$ are all rational (as they would be for any box spline with integer directions), then so will be the entries of the mesh function c of (16). Thus rounding errors incurred during the floating-point solving of (18) can be eliminated.

(20)Example (continued). In our bivariate example, we can think of c in (16) as an infinite *matrix*, on the index set $\{0,1,2,\ldots\} \times \{0,1,2,\ldots\}$ rather than the more standard index set $\mathbb{N} \times \mathbb{N}$. In terms of this mesh function, we can write g_α as

$$g_{i,j}(x,y) = \left([\![x]\!]^i, [\![x]\!]^{i-1}, \ldots, [\![x]\!]^0\right) c(0{:}i, 0{:}j) \left([\![y]\!]^j, [\![y]\!]^{j-1}, \ldots, [\![y]\!]^0\right)^T.$$

Here is the upper left corner of c for the ZP element:

$$c = \frac{1}{57600}\begin{pmatrix} 57600 & 86400 & 57600 & 21600 & 4800 & 720 & \cdot \\ 28800 & 43200 & 28800 & 10800 & 2400 & 360 & \cdot \\ 0 & 0 & 0 & 0 & 0 & 0 & \cdot \\ -2400 & -3600 & -2400 & -900 & -200 & -30 & \cdot \\ 0 & 0 & 0 & 0 & 0 & 0 & \cdot \\ 240 & 360 & 240 & 90 & 20 & 3 & \cdot \\ \cdot & \cdot & \cdot & \cdot & \cdot & \cdot & \cdot \end{pmatrix}$$

In this example, the linear functional μ has the specific form $\mu = (\eta \otimes \eta)\tau_{-1,-2}$, with $\tau_y f = f(\cdot + y)$ and $\eta := (\delta_0 + \delta_1)/2 : f \mapsto (f(0) + f(1))/2$ the linear functional whose Appell sequence (of univariate polynomials) is formed by the Euler polynomials $\mathcal{E}_j$. (In fact, probably the most efficient way to *define* the Euler polynomials is as the Appell polynomials for the linear functional $(\delta_0 + \delta_1)/2$.) Here, we employ **tensor-product** notation: If f and g are real-valued functions on $\mathbb{R}$, then $f \otimes g$ is the bivariate function given by

$$f \otimes g : \mathbb{R}^2 \to \mathbb{R} : x \mapsto f(x(1))g(x(2)).$$

Further, if η, θ are linear functionals, on $C(\mathbb{R})$ say, then $\eta \otimes \theta$ is the linear functional which takes the value $(\eta f)(\theta g)$ on $f \otimes g$. With this,

$$\mu D^{\ell,k}(\mathcal{E}_i(\cdot + 1) \otimes \mathcal{E}_j(\cdot + 2)) = (\eta \otimes \eta)(D^\ell \otimes D^k)(\mathcal{E}_i \otimes \mathcal{E}_j) = \delta_i(\ell)\delta_j(k),$$

therefore

$$g_{i,j} = \mathcal{E}_i(\cdot + 1) \otimes \mathcal{E}_j(\cdot + 2). \tag{21}$$

This shows that the earlier observation that $g_{1,1} = g_{1,0}g_{0,1}$ was no accident, since it shows that, more generally,

$$g_{i,j} = g_{i,0}g_{0,j}$$

for this example. It also explains why the matrix c above appears to be a rank-one matrix, i.e.,

$$c = \frac{1}{57600}(240\ \ 120\ \ 0\ \ -10\ \ 0\ \ 1\ \ \cdots)^T(240\ \ 360\ \ 240\ \ 90\ \ 20\ \ 3\ \ \cdots).$$

□

For the matrix $\Xi = \left[\begin{smallmatrix}1 & 1 & 0 & 0\\ 0 & 0 & 1 & 1\end{smallmatrix}\right]$, things are even simpler since now $\mu f = \sum M(-j)f(j) = f(-1,-1) = \delta_0\tau_{-1,-1}f$. Since the normalized powers form the univariate Appell sequence for the linear functional δ_0 of evaluation at 0, it follows that, for the present Ξ, the Appell polynomials are

$$g_\alpha = [\![\cdot + (1,1)]\!]^\alpha.$$

Choice of the quasi-interpolant functional. The linear functional

$$(22)\qquad \delta_0 \sum_{|\alpha|\le m(\Xi)} g_\alpha(0)\, D^\alpha : f \mapsto \sum_{|\alpha|\le m(\Xi)} g_\alpha(0)(D^\alpha f)(0)$$

obviously agrees on $\Pi_{m(\Xi)}$ with the linear functional λ_0 defined in (13). Hence, by (14)Theorem, Q_λ with λ given by (22) is certain to provide the **optimal** approximation order $m(\Xi)+1$ from S, as will any Q_λ (such as those generated in the examples below) whose λ agrees on $\Pi_{m(\Xi)}$ with (22).

(23)Example. We continue (20)Example. Since the columns of the direction matrix $\Xi = \left[\begin{smallmatrix}1 & 0 & 1 & -1\\ 0 & 1 & 1 & 1\end{smallmatrix}\right]$ for the ZP element are in general position, we have $D(\Xi) = \Pi_2$ (by (II.31)Proposition), hence $m(\Xi) = 2$ and, from (20), $g_{0,0}(0) = 1$, $g_{1,0}(0) = 1/2$, $g_{0,1}(0) = 3/2$, $g_{1,1}(0) = 3/4$ and $g_{2,0}(0) = 0$, $g_{0,2}(0) = 1$. Therefore, (22) gives in this case

$$\lambda f = f(0) + D^{1,0}f(0)/2 + 3D^{0,1}f(0)/2 + 3D^{1,1}f(0)/4 + D^{0,2}f(0),$$

hence the optimal approximation order is provided by

$$\sigma_h Q \sigma_{1/h} f = \sum_j M(\cdot/h - j)[f(hj) + h(D^{1,0}f)(hj)/2 + 3h(D^{0,1}f)(hj)/2$$
$$+3h^2(D^{1,1}f)(hj)/4 + h^2(D^{0,2}f)(hj)]. \square$$

If we use the max-norm on G, then point evaluation is continuous and we can write the linear functional (22) on $\Pi_{m(\Xi)}$ as a linear combination of evaluations at integer points near 0. In principle, we could express the linear functionals $\delta_0 D^\alpha$ restricted to $\Pi_{m(\Xi)}$ as linear combinations of point evaluations at the points $j \in \mathbb{Z}^s$, $|j| \le m(\Xi)$ (this is possible since these points are total for $\Pi_{m(\Xi)}$, i.e., the only $p \in \Pi_{m(\Xi)}$ vanishing at all these points is the zero polynomial), and then sum as in (22).

(24)Example. For the ZP element, this brute-force approach would give the linear functional $\lambda = \sum_j \delta_j w(j)$ whose non-zero weights are shown in (25)Figure(a) along with an outline of the support of the ZP element. (25)Figure(d) shows the weights w resulting from the same point pattern, but shifted closer to the center of the element. (25)Figure(c) shows the weights resulting from a centered and symmetric point pattern. There are only 5 (nonzero) weights; still, the corresponding λ matches λ_0 on all of Π_2.

By contrast, the choice $\lambda = \sum_{j\in J} \delta_j/4$, with $J := \{(0,1),(1,1),(0,2),(1,2)\}$ the four integer points in the interior of the support of the ZP element, will give $Q_\lambda = 1$ only on $\Pi_{1,1}$ and on no larger *translation-invariant* subspace of Π_2. (The qualifier 'translation-invariant' is important here since any λ whatsoever will agree with λ_0 on some subspace of Π_2 of codimension at most 1). However, if we add to these four points also the center of the ZP element, we obtain $\lambda = 2\delta_{(1,3)/2} - \sum_{j\in J}\delta_j/4$, i.e., the λ shown in (25)Figure(f), which does match λ_0 on all of Π_2.

Finally, (25)Figure(e) shows another way of augmenting J by just one point, an integer point this time, in order to obtain a fully matching λ. ◻

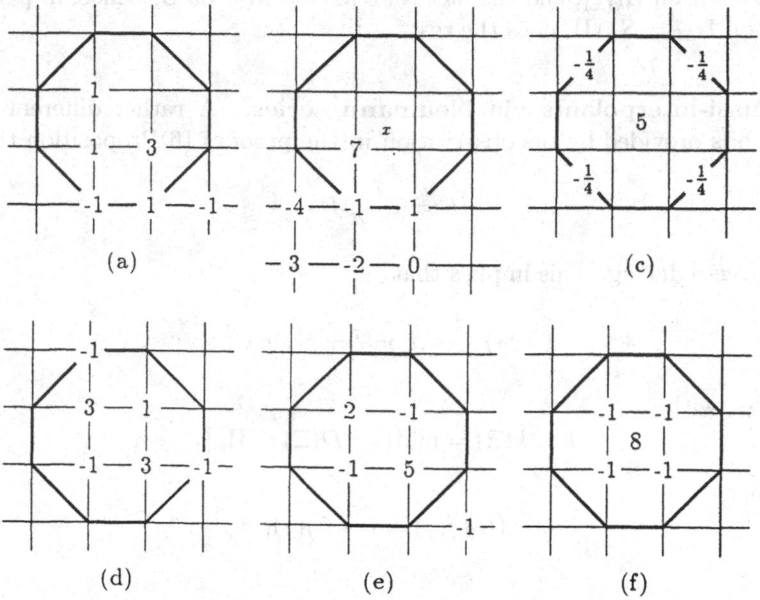

(25)Figure. Several λ which match λ_0 for the ZP element on $D(\Xi) = \Pi_2$. All weights are multiplied by 4.

Minimality. Already in the very simple setting of the ZP element, there are various simple questions as yet unanswered. Among these are: What is the mask with minimal norm? How small can one make the support of a mask? What is the relationship of a 'minimal' mask to Gauss quadrature?

If we want to have $Q = Q_\lambda$ reproduce all of Π_M, we have to choose λ so as to match λ_0 on all of Π_M. For this, recall from (II.46)Theorem that, for all $x \notin \Gamma(\Xi)$, point evaluations at the points in $\iota(x, \Xi)$ are total over $\Delta(\Xi)$, hence over Π_M since $\Pi_{M|} = D(\Xi)_| \subseteq \Delta(\Xi)$. This implies that we can always write λ_0 on Π_M as a linear combination of point evaluations at the points in $\iota(x, \Xi)$. Moreover, we cannot, in general, use fewer than $\#\iota(x, \Xi)$ points in case $\Delta(\Xi) = D(\Xi)_|$ since then $\#\iota(x, \Xi) = \dim D(\Xi)$, hence the points in $\iota(x, \Xi)$ are *minimally* total over $D(\Xi)$.

(26)Example. (25)Figure(b) shows one of the several possible $\lambda = \sum_{j \in \iota(x,\Xi)} \delta_j w(j)$ for $x := (1/2, 4/3)$ and the ZP element, hence incidentally indicates $\iota(x, \Xi)$ for this case. Note that $\#\iota(x, \Xi) = 7 > 6 = \dim D(\Xi)$ in this case, and that the additional freedom was used to make one of the weights 0. ◻

If we do not have Π_M available explicitly, we can still insure that Q_λ reproduces Π_M by ensuring that $\lambda = \lambda_0$ on $\Pi_{k(\Xi)}$. For, this ensures that $M_|{*}\nu{*} = 1$ on $\Pi_{k(\Xi)}$, and the fact that $M_|{*} = M{*}'$ on S, hence, in particular, on $\Pi_M = S \cap \Pi$, does the rest.

Quasi-interpolants via Neumann series. A rather different approach is provided by the observation in the proof of (6)Proposition that

$$U := 1 - M_|{*}$$

is degree-reducing. This implies that

$$(M_|{*})^{-1} = 1 + U + \ldots + U^k$$

on Π_M, with

$$k = k(\Xi) = \min\{r : D(\Xi) \subseteq \Pi_r\}.$$

Further,

$$(Uf)(j) = (v{*}f_|)(j),$$

with

$$v := \delta_0 - M_|$$

and δ_0 in this context the **unit mesh function**, i.e., $\delta_0(0) = 1$ and $\delta_0(j) = 0$ for $j \neq 0$. Hence, for $p \in \Pi_M$,

$$((M_|{*})^{-1}p)(0) = \sum_{\ell=0}^{k}(U^\ell p)(0) =: p^{[k]}(0)$$

with $p^{[k]}$ obtained inductively in the following computation:

$$(27) \qquad p^{[r]} := \begin{cases} p_|, & \text{if } r = 0; \\ p_| + v{*}p^{[r-1]}, & \text{if } r > 0. \end{cases}$$

This gives

$$\lambda_0 p = ((M_|{*})^{-1}p)(0) = \sum_{\mathbb{Z}^s} w(-j)p(j), \quad \text{all} \quad p \in \Pi_M$$

with the weight sequence w given by

$$w = \delta_0 + v + v{*}v + \ldots + \underbrace{v{*}\cdots{*}v}_{k \text{ factors}}.$$

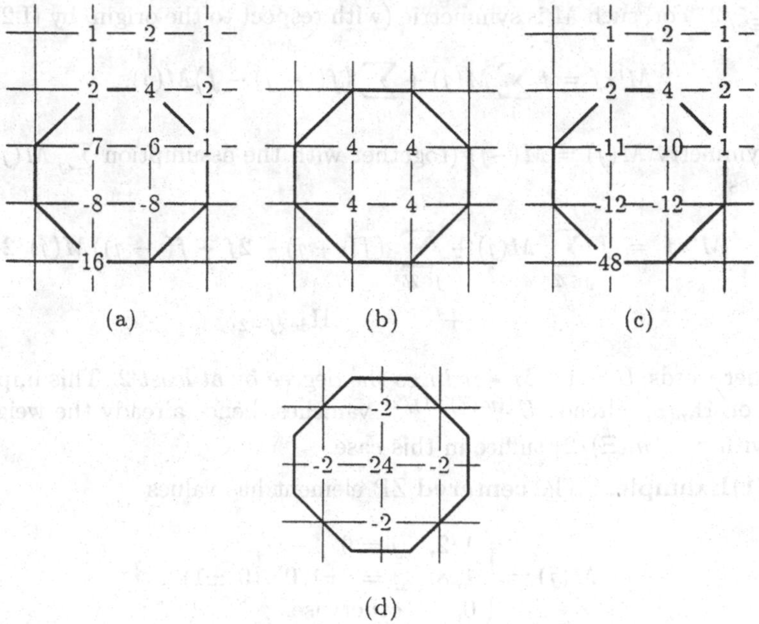

(28)Figure. (a) $v{*}v$, (b) $M_{|}$, (c) $w = \delta_0{+}v{+}v{*}v$ for the ZP element. (d) w for the centered ZP element. All weights are multiplied by 16.

Thus, the resulting quasi-interpolant functional

$$\lambda : f \mapsto \sum_{\mathbb{Z}^s} w(-j)f(j)$$

involves only finitely many mesh-values.

(29)Example. We continue to use the ZP element as an illustration. For the ZP element, $k(\Xi)$=2, so that $w = \delta_0 + v + v * v$. From (20) we have

$$v(j) = \begin{cases} 1, & \text{if } j = 0, \\ -1/4, & \text{if } j = (0,1),(1,1),(0,2),(1,2), \\ 0, & \text{otherwise.} \end{cases}$$

The non-trivial part of the resulting mesh functions $16v{*}v$ and $16w$ are shown in (28)Figure(a) and (c). Note that the support of λ is the negative of the support of w. □

If we are only after an optimal order quasi-interpolant, then we would be satisfied with matching λ_0 on $\Pi_{m(\Xi)}$, i.e., use the weights $p^{[m(\Xi)]}$ instead of the weights $p^{[k]}$. In fact, we can do even better than that by using the centered box spline $M = M_{\Xi}^{c} = M_{\Xi}(\cdot + c_{\Xi})$ instead of M_{Ξ}, with $c_{\Xi} =$

$\sum_{\xi\in\Xi}\xi/2$. For, such M is symmetric (with respect to the origin, by (I.22)). Since

$$M^{|}*f = f\sum_j M(j) + \sum_j (f(\cdot - j) - f)M(j),$$

the symmetry $M(j) = M(-j)$ (together with the assumption $\sum_j M(j) = 1$) gives

$$\begin{aligned} M^{|}*f &= f\sum_{j\in\mathbb{Z}^s} M(j) + \sum_{j\in\mathbb{Z}^s} (f(\cdot + j) - 2f + f(\cdot - j))M(j)/2 \\ &\in \quad f \quad + \quad \Pi_{\deg f - 2}. \end{aligned} \tag{30}$$

In other words, $U = 1 - M^{|}*$ *reduces the degree by at least 2.* This implies that, on $\Pi_{m(\Xi)}$, already $U^{\lfloor (m(\Xi)+1)/2\rfloor}$ vanishes, hence already the weights $p^{[r]}$ with $r = \lfloor m(\Xi)/2\rfloor$ suffice in this case.

(31)Example. The **centered** ZP element has values

$$M(j) = \begin{cases} 1/2, & j = 0 \\ 1/8, & j = (\pm 1, 0), (0, \pm 1) \\ 0, & \text{otherwise.} \end{cases}$$

Therefore, $v(0) = 1/2$ and $v(j) = -1/8$ for $j \in \pm\{(0,1),(1,0)\}$. Using only the weights $p^{[r]}$ with $r = \lfloor m(\Xi)/2\rfloor = 1$, we find that

$$\lambda_0 f = \lambda f := \frac{3}{2}f(0,0) - \frac{1}{8}(f(1,0) + f(-1,0) + f(0,1) + f(0,-1))$$

for $f \in \Pi_2$. This weight function is also shown in (28)Figure(d). It is comparable to the better choices shown in (25)Figure. But note that, of all the various representations $\sum_j w(j)\delta_j$ for λ_0 of the ZP element shown in (25)Figure and (28)Figure, (25)Figure(c) has the smallest 1-norm (i.e., the smallest norm as a linear functional on $C(\mathbb{R}^s)$). □

Quasi-interpolants via Fourier transform. It is also possible to show that $M*$ agrees on Π_M with $M^{|}*$. Thus, we could also construct suitable Q_λ by having λ match the linear functional

$$\delta_0\,(M*)^{-1} = \sum_\alpha h_\alpha(0)\delta_0 D^\alpha \tag{32}$$

on Π_M, with $(M*)^{-1}$ the inverse of $M*$ on Π, and with (h_α) the Appell polynomials for $\delta_0 M*$. Of course, necessarily $h_\alpha = g_\alpha$ for all $[]^\alpha \in \Pi_M$, so that there seems to be no gain here over the earlier approach. But, since the Fourier transform (cf. (I.17))

$$\widehat{M_\Xi}(y) = \prod_{\xi\in\Xi} \frac{1 - \exp(-i\xi y)}{i\xi y} \tag{33}$$

of the box spline $M = M_\Xi$ is so simple, the following alternative computation of the numbers $g_\alpha(0)$ may, at times, be more efficient than their computation via Appell polynomials.

We know from (I.5) that $M*[]^0 = 1$ for any box spline M, hence know, e.g. from (6) and (8), that $M*$ is invertible on Π. Further, we conclude from (7) that the Appell polynomials for $\mu = \delta_0\,(\nu*_{|\Pi})^{-1}$ are given by

$$h_\alpha = \sum_\beta []^{\alpha-\beta}[\![-iD]\!]^\beta \widehat{\nu}(0).$$

Since $M*\nu* = 1$ on Π implies that $1 - \widehat{M}\widehat{\nu}$ vanishes at 0 to any order, we conclude that the Appell polynomials for $\mu = \delta_0 M*$ are given by

$$h_\alpha = \sum_\beta []^{\alpha-\beta}[\![-iD]\!]^\beta (1/\widehat{M})(0).$$

In particular, the mesh function c of (16) satisfies

$$c(\beta) = [\![-iD]\!]^\beta (1/\widehat{M})(0). \tag{34}$$

In fact, once we bring in the Fourier transform, we can solve the problem of choosing ν so that $M^{|}*\nu* = 1$ on some Π_α directly by observing that this implies that $(\widehat{M^{|}}\widehat{\nu})(0) = 1$ α-fold, i.e.,

$$D^\beta \widehat{\nu}(0) = D^\beta (1/\widehat{M^{|}})(0) \qquad \text{all } \beta \le \alpha. \tag{35}$$

For this, note that $\widehat{M^{|}}$ is the **symbol** of M, i.e., the trigonometric polynomial

$$\widetilde{M} := \sum_j M(j)\exp(-ij(\,)). \tag{36}$$

(37) Example: the ZP element. For the ZP element, i.e., for $M = M_\Xi$ with $\Xi = \left[\begin{smallmatrix}1 & 0 & 1 & -1\\ 0 & 1 & 1 & 1\end{smallmatrix}\right]$, the symbol is the trigonometric polynomial

$$\begin{aligned} p(u,v) &= \exp(-i(u/2+3v/2))\;(1/4)(\textstyle\sum \exp(-i(\pm u/2 + \pm v/2))) \\ &= \exp(-i(u/2+3v/2))\;(\cos((u+v)/2) + \cos((u-v)/2))/2. \end{aligned} \tag{38}$$

Therefore, $c(0,0) = (1/p)(0,0) = 1$, and $c(1,0) = (-iD)^{1,0}(1/p)(0,0) = -1/(-iD)^{1,0}p(0,0) = -1/(-i(-i/2+0)) = 1/2$, and, finally, $c(0,1) = (-iD)^{0,1}(1/p)(0,0) = -1/(-iD)^{1,0}p(0,0) = -1/(-i(-i3/2+0)) = 3/2$, in agreement with (20)Example. Already the (hand) calculation of $c(1,1)$ requires greater fortitude. □

The efficient calculation of the first few numbers $[\![-D]\!]^\alpha(1/\widehat{\nu})(0)$ for given $\widehat{\nu}$ is arithmetically the same as the calculation of the first few entries of the mesh function c in (16) from (18). One can either proceed by induction, using the fact that a formula for $D^\alpha(1/\widehat{\nu})(0)$ involves all the numbers $D^\beta(1/\widehat{\nu})(0)$ for $\beta < \alpha$ (as well as the numbers $D^\beta\widehat{\nu}(0)$ for $\beta \le \alpha$), or else solve, for some α, the linear system for the numbers $D^\beta(1/\widehat{\nu})(0)$ for $\beta \le \alpha$ which results from the requirement that $\widehat{\nu}\cdot(1/\widehat{\nu}) = 1$ at 0 to all terms of order $\le \alpha$. Explicitly,

$$\delta_0(\beta) = D^\beta[\![]\!]^0(0) = \sum_{\gamma\le\beta}\Big([\![D]\!]^{\beta-\gamma}\widehat{\nu}(0)\Big)\,[\![D]\!]^\gamma(1/\widehat{\nu})(0), \qquad \beta \le \alpha,$$

by Leibniz' formula, thus providing a triangular system of equations for the unknowns $[\![D]\!]^\gamma(1/\widehat{\nu})(0)$, $\gamma \le \alpha$. For this, the information about ν needed are the numbers $D^\beta\widehat{\nu}(0)$ for $\beta \le \alpha$, and these might be more readily available for $\nu = M_\Xi$ than are the numbers $\nu(j)$, $j \in \mathbb{Z}^s$, which the earlier, direct approach requires.

Notes. The basic results on approximation order ((1)Proposition and (4)Proposition) are taken from [de Boor, Höllig'82b], but are extended here to all $\mathbf{L}_p$.

The quasi-interpolant construct originated in Finite Element Analysis (see, e.g., [Strang, Fix'73a]) (but was already used in [de Boor'68b] to determine lower bounds for the approximation order of general (univariate) spline spaces). The abstract construction of a *quasi-interpolant* for a box spline space can be found in [de Boor, Höllig'82b]. The explicit constructions of suitable linear functionals reported here are taken from the following sources: The Fourier Transform approach is due to Strang & Fix [Strang, Fix'73b] (and to the sources quoted therein), and was specialized to box splines in [Dahmen, Micchelli'84b,'84c]. The claim that $M* = M^{|}*$ on Π_M is proved in [de Boor, Ron'92c]. The Neumann series approach is due to Chui & Diamond [Chui, Diamond'87]. In particular, the possibility of using symmetry to halve the number of necessary terms is rightfully stressed there. Finally, the Appell sequence approach is taken from [de Boor'87c], as are the general considerations concerning *semi-discrete convolutions.* That paper was strongly influenced by [Chui, Jetter, Ward'87], where it is proved that cardinal spline interpolation provides optimal approximation order in $\mathbf{L}_2$. A version of the multivariate *Marsden identity* can already be found in [Chui, Lai'87a].

IV

Cardinal interpolation & difference equations

In this chapter, we consider cardinal interpolation, i.e., the interpolation to data on the integer mesh, from a space of cardinal splines. This problem is intimately connected to the solution of a multivariate difference equation, and the qualitative properties of the interpolation, its correctness or singularity, are determined by the symbol of the difference equation.

Correct and singular cardinal interpolation. A **cardinal interpolant** to a sequence f on $\mathbb{Z}^s$ is a cardinal spline

$$\sum_{j\in\mathbb{Z}^s} M(\cdot - j)a(j) = M{*}a \in S$$

which matches f on $\mathbb{Z}^s$, i.e., whose coefficients a satisfy the difference equation

$$(M{*}a)(k) = \sum_{j\in\mathbb{Z}^s} M(k-j)a(j) = f(k),\ k \in \mathbb{Z}^s. \tag{1}$$

Since, in general, the difference equation (1) does not have a unique solution, additional conditions are required for the definition of the cardinal interpolant. We say that cardinal interpolation with M is **correct** if for any *bounded* f there exists a unique *bounded* solution a of (1), i.e., if the mapping

$$M_{|}{*} : a \mapsto \sum_{j\in\mathbb{Z}^s} M(\cdot - j)_{|}a(j)$$

carries $\ell_\infty(\mathbb{Z}^s)$ one-one onto $\ell_\infty(\mathbb{Z}^s)$. If, on the other hand, there exists a bounded **null-solution**, i.e., a (bounded) nontrivial sequence a with $M_|{*}a = 0$, then we say that cardinal interpolation with M is **singular**. (12)Theorem shows that these are the only two possibilities.

In the univariate case it is known that cardinal interpolation by the translated B-spline $M_n(\cdot - \alpha)$ is correct unless $\alpha = (n+1)/2 \bmod \mathbb{Z}$. In particular, for n odd, the interpolation is singular for $\alpha = 0$. This has immediate consequences for higher dimensions, as the next example shows, but it is not indicative of the multivariate theory where correctness is rare and a richer theory for the singular case exists.

In this example (and others in this chapter and the next), we use bivariate box splines on the three-direction mesh. These, to recall from (II.44)Example, are box splines for the direction matrix

$$\Xi = T_r = \left[\begin{smallmatrix}1 & 0 & 1\\ 0 & 1 & 1\end{smallmatrix}\right]_r,$$

with the triple r indicating the multiplicities with which each of the three directions $\mathbf{i}_1, \mathbf{i}_2, \mathbf{i}_1{+}\mathbf{i}_2$ occurs in Ξ. In analogy with the notation M_n for the univariate cardinal spline, we write correspondingly

$$M_r := M_{T_r}.$$

We use the analogous notation for box splines on the four-direction mesh (cf. (I.39)), i.e., define

$$M_r := M_{V_r}$$

where now r is an integer 4-vector and

$$V := \left[\begin{smallmatrix}1 & 0 & 1 & -1\\ 0 & 1 & 1 & 1\end{smallmatrix}\right].$$

(2)Example: univariate and bivariate quadratic box splines. Consider univariate cardinal interpolation with M the translated quadratic B-spline $M_3(\cdot - \alpha)$. For $\alpha = -3/2$, the difference equation (1) becomes

$$\text{(3)} \qquad \frac{1}{8}a(k-1) + \frac{3}{4}a(k) + \frac{1}{8}a(k+1) = f(k),$$

while, for $\alpha = 0$,

$$\text{(4)} \qquad \frac{1}{2}a(k-2) + \frac{1}{2}a(k-1) = f(k).$$

Cardinal interpolation with M the centered B-spline $M_3(\cdot + 3/2)$ is correct since, in this case, $M_|{*}$ is represented by a diagonally dominant matrix. However, for the forward B-spline M_3, interpolation at the knots is singular, since (4) has the bounded null-solution $j \mapsto (-1)^j$. This singularity persists

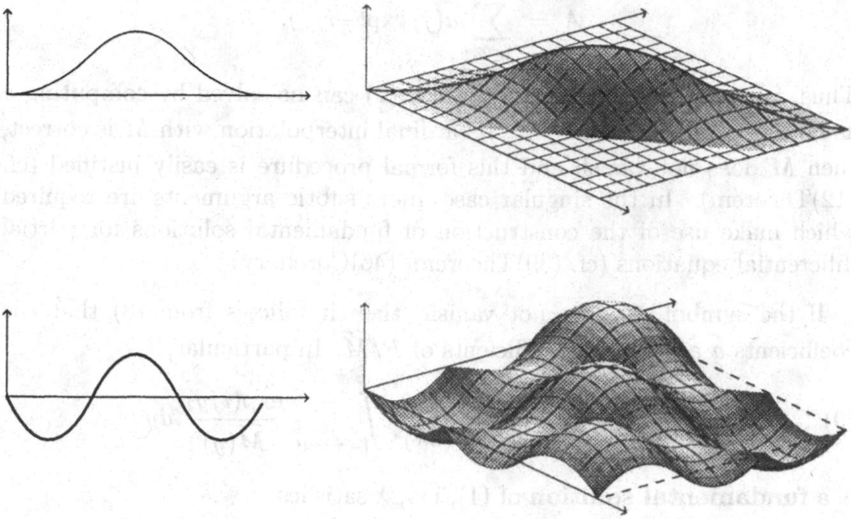

(5)Figure. The univariate quadratic B-spline and the bivariate quadratic box spline $M_{1,1,2}$ with corresponding null solutions.

for the quadratic box spline $M_{1,1,2}$ in $\mathbb{R}^2$. The restriction of $M_{1,1,2}$ to the line $x(1) - x(2) = 0$ equals M_3 and satisfies the corresponding difference equation

$$\frac{1}{2}a(k - 2(1,1)) + \frac{1}{2}a(k - (1,1)) = f(k), \qquad k \in \mathbb{Z}^2. \tag{6}$$

One of the (infinitely many linearly independent) bounded null-solutions for (6) is $a : j \mapsto \cos((j(1) + j(2))\pi/2)$. (5)Figure illustrates the splines of this example. ▫

Symbol and fundamental solution. The **symbol** of the difference equation (1) is, by definition, the 2π-periodic function

$$\widetilde{M}(y) = \sum_{j \in \mathbb{Z}^s} M(-j)\exp(ijy), \quad \text{for } y \in \mathbb{R}^s. \tag{7}$$

The symbol is of interest since linear combinations of solutions of the form $j \mapsto z^j$ are dense in the set of all null-solutions. In particular, (1) has bounded null-solutions if and only if $\widetilde{M}$ vanishes somewhere (this is (12)Theorem).

Multiplying both sides of (1) by $\exp(-iky)$ and summing over k yields

$$\widetilde{M}A = F, \tag{8}$$

where A and F are the conjugate Fourier series with coefficients a and f respectively. E.g.,

$$A = \sum_{j \in \mathbb{Z}^s} a(j) \exp(-ij\cdot).$$

Thus, formally, the difference equation (1) can be solved by computing a as Fourier coefficients of $F/\widetilde{M}$. If cardinal interpolation with M is correct, then $\widetilde{M}$ does not vanish and this formal procedure is easily justified (cf. (12)Theorem). In the singular case, more subtle arguments are required which make use of the construction of fundamental solutions for partial differential equations (cf. (39)Theorem, (46)Corollary).

If the symbol $\widetilde{M}$ does not vanish, then it follows from (8) that the coefficients a are Fourier coefficients of $F/\widetilde{M}$. In particular,

$$\lambda : j \mapsto (1/\widetilde{M})^{\vee}(j) := \frac{1}{(2\pi)^s} \int_{[-\pi..\pi]^s} \frac{\exp(ijy)}{\widetilde{M}(y)}\, dy \tag{9}$$

is a **fundamental solution** of (1), i.e., λ satisfies

$$\sum_{j \in \mathbb{Z}^s} M(k-j)\lambda(j) = \delta(k) := \begin{cases} 1, & \text{if } k = 0; \\ 0, & \text{otherwise.} \end{cases} \tag{10}$$

With

$$L := M*\lambda \in S$$

the corresponding **fundamental function**, the cardinal interpolant to a sequence f on $\mathbb{Z}^s$ can be written in **Lagrange form**,

$$\mathcal{L}f := L*f = \sum_{j \in \mathbb{Z}^s} L(\cdot - j) f(j), \tag{11}$$

provided that the sum converges.

(12)Theorem. *Let M be any (translate of a) box spline. Then, cardinal interpolation with M is correct if and only if the symbol $\widetilde{M}$ does not vanish. In this case,*

$$|L(x)| \le \text{const} \exp(-|x|/\text{const}), \tag{13}$$

for some positive constants const *and, for any $p \in [1..\infty]$ and any $f \in \ell_p(\mathbb{Z}^s)$,*

$$\|\mathcal{L}f\|_p \le \text{const} \|f\|_p. \tag{14}$$

Proof. If $\widetilde{M}(y_0) = 0$, then the sequence $j \mapsto \exp(iy_0 j)$ is in $\ker M_{|}*$, i.e., (1) has a bounded null-solution.

On the other hand, if $\widetilde{M}$ does not vanish, then $1/\widetilde{M}$ is analytic in a neighborhood of $[-\pi\,..\,\pi]^s$ in $\mathbb{C}^s$. From Cauchy's integral theorem and the periodicity of $\exp(ij\cdot)/\widetilde{M}$, it follows that the integral in (9) may be replaced by the integral

$$\int_{\mathcal{C}} \frac{\exp(ijy)}{\widetilde{M}(y)}\,dy$$

where the contour $\mathcal{C}$ is given by

$$\mathcal{C} = [-\pi\,..\,\pi]^s + (\varepsilon \operatorname{sgn} j)i = \mathop{\times}_{\nu} [-\pi + (\varepsilon \operatorname{sgn} j_\nu)i\,..\,\pi + (\varepsilon \operatorname{sgn} j_\nu)i]$$

for some $\varepsilon > 0$. Therefore,

$$|(1/\widetilde{M})^\vee(j)| \le \|1/\widetilde{M}_{|\mathcal{C}}\|_\infty \exp(-\varepsilon|j|), \qquad j \in \mathbb{Z}^s,$$

which implies (13) since M has compact support.

For the proof of (14) we recall Hölder's inequality,

$$(15) \qquad |\sum_{j\in\mathbb{Z}^s} b(j)c(j)| \le \|b\|_p \|c\|_{p'}, \quad 1/p + 1/p' = 1.$$

Since

$$|L(x-j)| \le \operatorname{const}\exp(-|k-j|/\operatorname{const}) =: e(k-j) \quad \forall x \in \square + k,$$

applying (15) with $b := e(k-\cdot)^{1/p} f$, $c := e(k-\cdot)^{1/p'}$ yields the second inequality in the following familiar way:

$$\begin{aligned}
\|\mathcal{L}f\|_p &\le \sum_{k\in\mathbb{Z}^s} \Big(\sum_{j\in\mathbb{Z}^s} e(k-j)|f(j)|\Big)^p \\
&\le \sum_k \Big(\sum_j e(k-j)|f(j)|^p\Big)\Big(\sum_j e(k-j)\Big)^{p/p'} \\
&\le \sum_j |f(j)|^p \sum_k e(k-j)\|e\|_1^{p/p'} \le \operatorname{const}\|f\|_p^p.
\end{aligned}$$

□

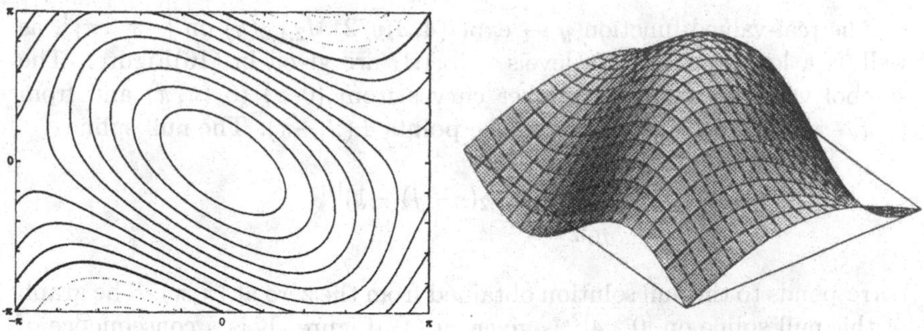

(16)Figure. Level curves and plot of $\exp(i(4,3)/2\cdot)\widetilde{M}_{2,1,2}$.

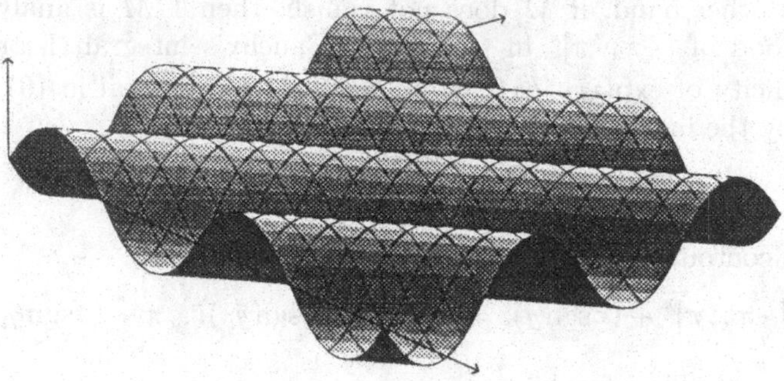

(17)Figure. A null spline for $M_{2,1,2}$.

(18)Example: three-direction mesh. The symbols for the three-direction box splines $M_{2,1,2}$, $M_{2,2,2}$ and $M_{4,2,2}$ in $\mathbb{R}^2$ are

$$\begin{aligned}\widetilde{M}_{2,1,2}(y) &= \Big(\exp(-i(1,1)y) + 2\exp(-i(2,1)y) + 2\exp(-i(2,2)y) \\ &\qquad + \exp(-i(3,2)y)\Big)/6 \\ &= \exp(-i(4,3)y/2)\Big(2\cos((0,1)y/2) + \cos((2,1)y/2)\Big)/3, \\ \widetilde{M}_{2,2,2}(y) &= \exp(-i(2,2)y)\Big(3 + \cos((1,0)y) + \cos((0,1)y) \\ &\qquad + \cos((1,1)y)\Big)/6, \\ \widetilde{M}_{4,2,2}(y) &= \exp(-i(3,2)y)\Big(71 + 48\cos((1,0)y) \\ &\qquad + 28\cos((0,1)y) + 28\cos((1,1)y) \\ &\qquad + 2\cos((1,-1)y) + \cos((2,0)y) + 2\cos((2,1)y)\Big)/180,\end{aligned}$$

respectively.

The real-valued function $y \mapsto \exp(i(4,3)y/2)\widetilde{M}_{2,1,2}(y)$ on $[-\pi \mathinner{..} \pi]^2$, as well as a level plot for the levels $-.3$:.1:1, are given in (16)Figure. The symbol vanishes along the darker curves from $(0,\pi)$ to (π,π) and from $(-\pi,-\pi)$ to $(0,-\pi)$ as well as at the points $\pm(\pi,-\pi)$. The null spline

$$\sum_{j\in\mathbb{Z}^2} M_{2,1,2}(\cdot - j)(-1)^{|j|}$$

corresponds to the null solution obtained from the zero at (π,π). The graph of this null spline on $[0 \mathinner{..} 4]^2$ is given in (17)Figure. It is a consequence of (32)Proposition that the symbols $\widetilde{M}_{2,2,2}$ and $\widetilde{M}_{4,2,2}$ are never zero. Therefore, the fundamental functions for cardinal interpolation by $M_{2,2,2}$ and

$M_{4,2,2}$ have the exponential decay of (12)Theorem. These fundamental functions are shown in (19)Figure to scale.

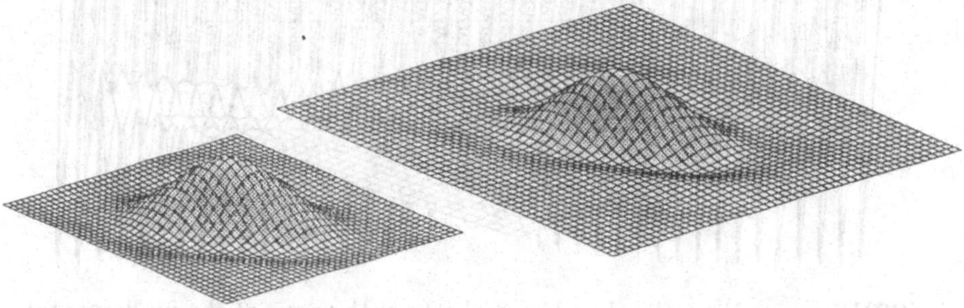

(19)Figure. The fundamental spline for $M_{2,2,2}$ on $[-2\,..\,2]^2$, and for $M_{4,2,2}$ on $[-3\,..\,3]^2$.

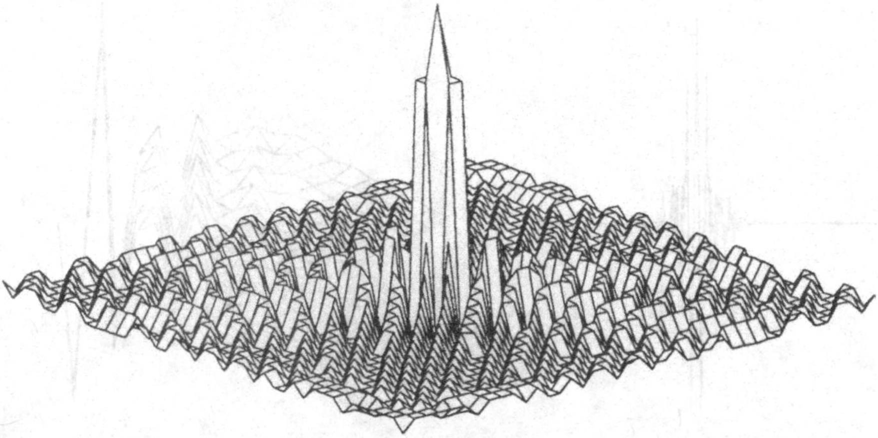

(20)Figure. Signed logarithm of the fundamental solution for (21).

The difference equation corresponding to $\widetilde{M}_{2,2,2}$ is

(21)

$$\tfrac{1}{2}a(k-(2,2))+\tfrac{1}{12}\{a(k-(1,1))+a(k-(1,2))+a(k-(2,1)) \\ +a(k-(2,3))+a(k-(3,2))+a(k-(3,3))\}=\delta(k).$$

The values, $\lambda(j-(2,2))$, $j\in[-20\,..\,20]^2$, for the fundamental solution of this equation exhibit the expected exponential decay and an interesting sign pattern. This is illustrated in (20)Figure where the mesh function $\operatorname{sgn}(\lambda)/\log(|\lambda|)$ is shown. The sign patterns follow elliptical contours around the diagonal.

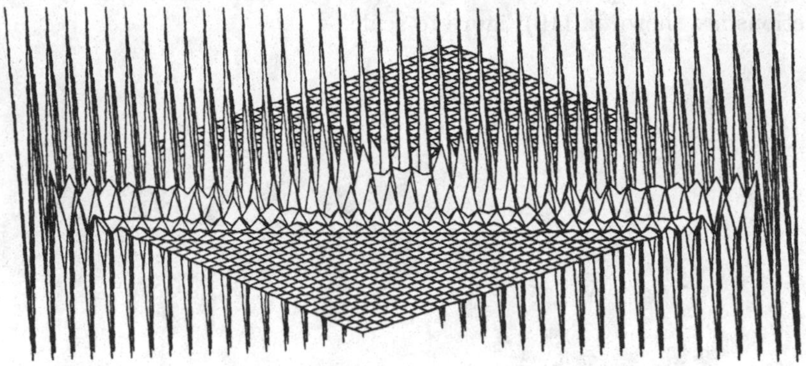

(22)Figure. Normalized scalar products of the rows of the fundamental solution matrix for (21).

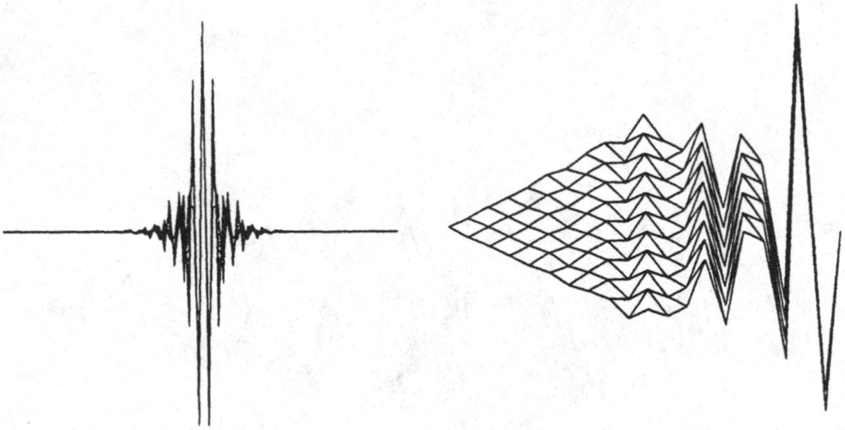

(23)Figure. Regularity of the sign and exponential decay of the scalar products for the fundamental solution matrix of (21).

The shifted λ obtained above is symmetric, as can be seen from the difference equation or from the equation for $\underset{\sim}{M}_{2,2,2}$. The corresponding matrix of the normalized scalar products of the rows of this matrix, i.e., the cosines of the angles between the row vectors, also tells us something about the regular behavior of these coefficients and their decay. (22)Figure shows these scalar products as a mesh function. The regularity of the sign pattern and the exponential decay of the normalized scalar products is evidenced by this figure as well as by the figures in (23)Figure where there is a view of the mesh function straight down the diagonal as well as a piece of it ten mesh points away from the diagonal. ☐

Correctness and linear independence. If the shifts of the box spline are linearly dependent, then, by (II.57)Theorem(i,vi) and (II.49)Theorem, the kernel of $M^\star$ contains a bounded exponential, i.e.,

$$\sum_{j\in\mathbb{Z}^s} M(\cdot - j)\exp(2\pi i\zeta j) = 0$$

for some $\zeta \in [0\,..\,1)^s\backslash 0$. This implies that $\widetilde{M}(2\pi\zeta) = 0$. Therefore, linear independence of the shifts of the box splines is necessary for the correctness of cardinal interpolation. On the other hand, it is not sufficient. This is already evident from the three-direction box splines $M_{1,1,2}$ and $M_{2,1,2}$.

Since the linear independence of the shifts is invariant under translation of the basic function, it is conceivable that an appropriate translate of the box spline could lead to correct interpolation. To explore this possibility, it is natural from symmetry considerations to first translate the origin to the center of the support of M_Ξ,

$$c_\Xi = \sum_{\xi\in\Xi} \xi/2.$$

Recall from Chapter I that the *centered box spline*, defined in (I.21) as $M_\Xi^c = M_\Xi(\cdot + c_\Xi)$, is a.e. symmetric with respect to the origin, and has the *real-valued* Fourier transform,

(24) $$\widehat{M_\Xi^c}(y) = \prod_{\xi\in\Xi} \operatorname{sinc}(y\xi).$$

Further from (I.22), the centered box spline is unchanged when $\xi \in \Xi$ is replaced by $-\xi$.

For $c_\Xi \in \mathbb{Z}^s$, cardinal interpolation with M_Ξ is correct if and only if it is so with M_Ξ^c. However, if $c_\Xi \notin \mathbb{Z}^s$, then cardinal interpolation with M_Ξ is singular. To see this, let $\widetilde{M_\alpha^c}$ be the symbol for the centered box spline translated by α; i.e.,

$$\widetilde{M_\alpha^c}(y) := \sum_{j\in\mathbb{Z}^s} M_\Xi^c(j-\alpha)\exp(-ijy).$$

We write, more simply, $\widetilde{M^c}$ when $\alpha = 0$. For any $\beta \in \mathbb{Z}^s$, $\widetilde{M^c_{-\beta/2}}(y) = \widetilde{M^c_{\beta/2}}(y)\exp(i\beta y)$, while $\widetilde{M_\alpha^c}(-y) = \widetilde{M^c_{-\alpha}}(y)$ a.e., because of the a.e. symmetry of M^c about the origin. Therefore,

$$\widetilde{M^c_{\beta/2}}(y)\exp(i\beta y) = \widetilde{M^c_{\beta/2}}(-y) \ \ a.e.$$

In particular, if M_Ξ is continuous and some component, $\beta(\nu)$, is odd, then $\widetilde{M^c_{\beta/2}}(\pi \mathbf{i}_\nu) = 0$. Taking $\beta = 2c_\Xi$, this shows:

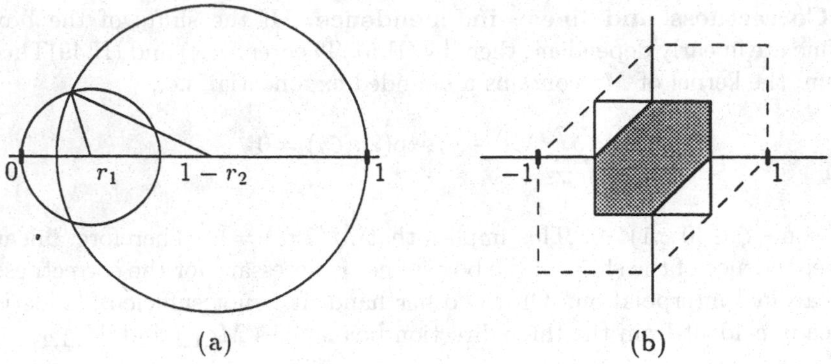

(27)Figure. The two circles have no point in common, i.e., cardinal interpolation with $M^c_{1,1,1}(\cdot - \alpha)$ is correct, if and only if α is in the (open) shaded region.

(25)Proposition. *Cardinal interpolation for a continuous M_Ξ is singular if $c_\Xi \notin \mathbb{Z}^s$.*

(26)Example: centered hat function. Consider the translates of the centered hat function, i.e., the bivariate centered three-direction box spline

$$M_\alpha := M^c_{1,1,1}(\cdot - \alpha)$$

with $\alpha \in [-1/2\,..\,1/2]^2$. Since the interior of $\alpha + T\sqcup$ contains at most three integer points, there are at most three nonzero coefficients, $M_\alpha(0)$, $M_\alpha(j_1) =: r_1$, $M_\alpha(j_2) =: r_2$, in $\widetilde{M}_\alpha$. Hence

$$\widetilde{M}_\alpha(y) = M_\alpha(0) + M_\alpha(j_1)\exp(-ij_1y) + M_\alpha(j_2)\exp(-ij_2y).$$

Since $M_\alpha(0) + M_\alpha(j_1) + M_\alpha(j_2) = 1$, and since j_1, j_2 are linearly independent, the equation $\widetilde{M}_\alpha(y) = 0$ is equivalent to the equation

$$r_1 - r_1z_1 = 1 - r_2 + r_2z_2$$

for the complex numbers z_1, z_2 of absolute value 1 (since the linear independence of j_1, j_2 implies that the equations $z_k = \exp(-ij_ky)$, $k = 1, 2$, can be solved for y for arbitrary such z_1, z_2). The equation for the z_k is one for the intersection of the two circles with centers r_1, $1 - r_2$, and radii r_1, r_2 respectively. These circles (see (27)Figure(a)) have a common point if and only if $2r_1 \geq 1 - 2r_2$, i.e., $2r_1 + 2r_2 \geq 1$, which, in view of the fact that $M_\alpha(0) + r_1 + r_2 = 1$ and $M_\alpha(0) = M^c_{1,1,1}(-\alpha)$, is equivalent to $M^c_{1,1,1}(-\alpha) \leq 1/2$. Since $M^c_{1,1,1}$ is piecewise linear on its support and $M^c_{1,1,1}(0) = 1$, cardinal interpolation with M_α, $\alpha \in [-1/2\,..\,1/2]^2$, is correct (equivalently, $M^c_{1,1,1}(-\alpha) > 1/2$) if and only if

$$\alpha \in J := \{\alpha : |\alpha(1)| < 1/2,\ |\alpha(2)| < 1/2,\ |\alpha(1) - \alpha(2)| < 1/2\}$$

(see (27)Figure (b)). □

Cardinal interpolation with any centered univariate box spline is correct. While this does not remain true in several variables (see, e.g., (49)Example), a simple sufficient condition for the correctness of cardinal interpolation can be given. For this we need

(28)Proposition. *The shifts of M are linearly independent (equivalently, $\Xi \in \mathbb{Z}^{s\times n}$ is unimodular) if and only if for every $y \in [0\,..\,2\pi)^s$, there exists $j \in \mathbb{Z}^s$ such that $\widehat{M}(2\pi j + y) \neq 0$.*

Proof. Since Ξ and $\Xi \cup \Xi$ are simultaneously unimodular (or not) and $\widehat{M}_{\Xi\cup\Xi} = \widehat{M}_\Xi^2$, the validity of the proposition for Ξ is equivalent to its validity for $\Xi \cup \Xi$. Hence (replacing Ξ with $\Xi \cup \Xi$ if necessary to ensure a continuous M), we may apply (see the Notes for a justification) the Poisson summation formula,

$$\sum_{j\in\mathbb{Z}^s} f(j) = \sum_{j\in\mathbb{Z}^s} \hat{f}(2\pi j), \tag{29}$$

with $f = M(x - \cdot)\exp(iy\cdot)$, to obtain

$$\sum_{j\in\mathbb{Z}^s} M(x-j)\exp(ijy) = \sum_{j\in\mathbb{Z}^s} \widehat{M}(y-2\pi j)\exp(ix(y-2\pi j)).$$

Hence,

$$\widehat{M}(2\pi j + y) = 0, \quad \text{for some} \quad y \in [0\,..\,2\pi)^s \quad \text{and all} \quad j \in \mathbb{Z}^s, \tag{30}$$

is equivalent to having $\exp(iy\cdot)$ in $\ker M^\star$, and having this for some $y \in [0\,..\,2\pi)^s$ is, by the equivalence (i)$\Longleftrightarrow$(vi) of (II.57)Theorem and the characterization of $\Delta(\Xi)$ given in (II.49)Theorem, equivalent to the shifts of M being linearly dependent. □

Applying the Poisson summation formula again, this time with $f := \exp(-iy\cdot)M$ and under the assumption that M is continuous, the symbol of M can be written in the form

$$\widetilde{M}(y) = \sum_{j\in\mathbb{Z}^s} \widehat{M}(2\pi j + y). \tag{31}$$

Therefore, if M is continuous and $\widehat{M}$ is nonnegative, then cardinal interpolation with M is not correct if and only if (30) holds. In particular:

(32)Proposition. *If $\widehat{M_\Xi^c}$ is nonnegative (and therefore M_Ξ^c is continuous), then cardinal interpolation with M_Ξ^c is correct if and only if its shifts are linearly independent.*

(33)Example. The Fourier transform of M_Ξ^c is nonnegative if, e.g., the directions in Ξ occur with even multiplicities. For example, recall from (II.29)Example that, up to symmetries, the only bivariate box splines whose shifts are linearly independent are those involving only the directions $\{\mathbf{i}_1, \mathbf{i}_2, \mathbf{i}_1 + \mathbf{i}_2\}$. Thus, up to symmetry and translation, the only bivariate box splines satisfying both conditions of (32)Proposition are the three-direction box splines M_r^c, with each $r(\nu)$, $\nu = 1, 2, 3$, even (and at most one $r(\nu) = 0$).

There is a little more freedom in three dimensions. Up to the signs of the columns and unitary changes of variable, only the centered box splines for matrices of the form

$$\begin{bmatrix} 1 & 0 & 0 & 1 & 0 & 1 \\ 0 & 1 & 0 & 1 & 1 & 0 \\ 0 & 0 & 1 & 1 & 1 & 1 \end{bmatrix}_r \quad \text{and} \quad \begin{bmatrix} 1 & 0 & 0 & 1 & 0 & -1 \\ 0 & 1 & 0 & 1 & 1 & 0 \\ 0 & 0 & 1 & 0 & 1 & 1 \end{bmatrix}_r$$

with even multiplicities $r(\nu)$ (and at most three $r(\nu) = 0$), $\nu = 1, \ldots, 6$, satisfy (32)Proposition. □

For any matrix $Q \in \mathbb{Z}^{s \times s}$ with $|\det Q| = 1$, we have $(M * a) \circ Q = (M \circ Q) * (a \circ Q)$. Therefore, the question of correctness of cardinal interpolation is equivalent for M and $M \circ Q$. In particular, it is equivalent for M_Ξ and $M_{Q\Xi}$, by the relation (I.23). Using (I.23) directly in the definition of the symbol, we obtain

$$\text{(34)} \qquad \widetilde{M_\Xi}(Q^T y) = \widetilde{M_{Q\Xi}}(y), \qquad \text{if } |\det Q| = 1,$$

with Q^T the transpose of Q. Since $Qc_\Xi = c_{Q\Xi}$, formula (34) applies equally well to centered box splines and provides symmetries that are useful in calculations.

(35)Example. For some insight into cases when (32)Proposition does not apply, it is instructive to continue (33)Example for the centered three-direction box spline with *odd* multiplicities $r(\nu)$, $\nu = 1, 2, 3$. The relation (34) will be used to reduce the domain that must be considered. With $Q = -\mathbb{1}$, we find that $\widetilde{M}(2\pi v) = \widetilde{M}(-2\pi v)$. The matrices

$$Q_1 = \begin{bmatrix} -1 & 0 \\ 1 & 1 \end{bmatrix} \quad \text{and} \quad Q_2 = \begin{bmatrix} 1 & 1 \\ 0 & -1 \end{bmatrix}$$

map the cones (see (38)Figure)

$$C_1 := \{y : y(1) + y(2) \geq 0,\ y(1) < 0\}$$

and

$$C_2 := \{y : y(1) + y(2) \geq 0,\ y(2) < 0\},$$

respectively, into

$$C := [0\,..\,\infty)^2.$$

Moreover, $Q_\mu^T T_r$ is T_{r_μ} (up to the signs of the columns), where r_μ is a permutation of r (cf. (V.22)Example). Therefore, it suffices to prove that $\widetilde{M}(2\pi v) > 0$ on $[0\,..\,1/2]^2$ for all odd multiplicities $r(\nu)$, $\nu = 1, 2, 3$.

With the aid of (24), the individual terms in the Poisson expansion (31) for $\widetilde{M}(2\pi v)$ may be written as

$$\widehat{M}(2\pi(j+v)) = (-1)^{\varepsilon(j)}\widehat{M}(2\pi v)G_j(v)$$

where $\varepsilon(j) := (r(1) + r(3))j(1) + (r(2) + r(3))j(2)$ is even and

$$G_j(v) := \left(\frac{v(1)}{j(1)+v(1)}\right)^{r(1)} \left(\frac{v(2)}{j(2)+v(2)}\right)^{r(2)} \left(\frac{v(1)+v(2)}{(j+v)(1) + (j+v)(2)}\right)^{r(3)}.$$

To show the positivity of $\widetilde{M}(2\pi v)$, we choose an appropriate pairing of the terms $G_j(v)$ through the origin by using the three cones C, C_1 and C_2. Consider the pairings

(36) $$G_j(v) + G_{-j-(1,1)}(v) \quad \text{for} \quad j \in C$$

and

(37) $$\begin{aligned} &G_{-j-\mathrm{i}_1}(v) + G_j(v) \quad \text{or} \quad G_{-j-\mathrm{i}_1}(v) + G_{j+\mathrm{i}_2}(v) \quad \text{for } j \in C_1, \\ &G_{-j-\mathrm{i}_2}(v) + G_j(v) \quad \text{or} \quad G_{-j-\mathrm{i}_2}(v) + G_{j+\mathrm{i}_1}(v) \quad \text{for } j \in C_2. \end{aligned}$$

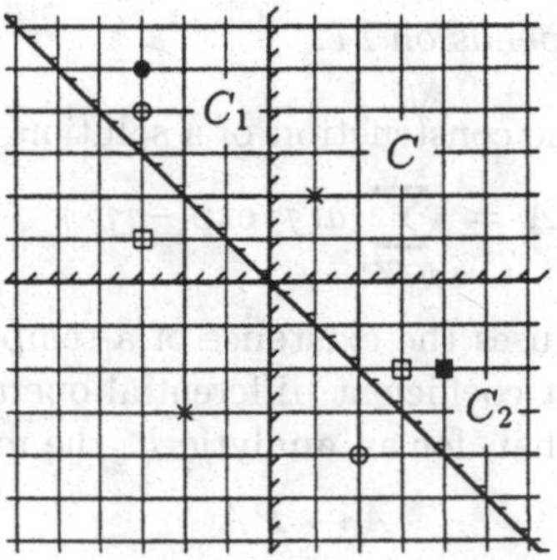

(38)Figure. The pairing of terms G_j through the origin.

In each case the first summand is positive, the second is negative and, for appropriate v, each factor of the first dominates the corresponding factor of the second, hence the sums are nonnegative. The comparison of factors is valid for all $v \in [0\,..\,1/2]^2$ for (36) and the choices on the right in (37), but is only true for the left hand pairing in (37) if v also satisfies

$v(1)+v(2)\geq 1/2$. In the latter case, all terms are taken into account and the sum is nonnegative.

When $v(1)+v(2)<1/2$, the right hand pairing in (37) is used. In this case, the unpaired terms $j(1)+j(2)=0$, $j\neq 0$, have a sum which is bounded below by the sum for $r(\nu)=1$, $\nu=1,2,3$. The latter sum may be rewritten and estimated as follows

$$-2v(1)v(2)\sum_{\ell>0}\frac{\ell^2-v(1)v(2)}{(\ell^2-v(1)^2)(\ell^2-v(2)^2)}\geq -2\frac{1}{16}\sum_{\ell>0}\frac{1}{\ell^2-(1/2)^2}=-\frac{1}{4},$$

when $0\leq \max\{v(1),v(2)\}\leq v(1)+v(2)\leq 1/2$. Then this sum may be combined with the $j=0$ term of (36) which is certainly $>2/3$ when $v(1)+v(2)\leq 1/2$, hence the sum is nonnegative in this case as well.

Finally, (36) can be zero for all $j\in C$ only for $v=0$ or $v=(1/2,1/2)$. But $\widetilde{M}(0)=1$ and the term $\widehat{M}(2\pi v)(G_{-i_2}(v)+G_{-i_1+i_2}(v))>0$ at $v=(1/2,1/2)$. Thus, the positivity of $\widetilde{M}$ is established. □

Singular cardinal interpolation. If the symbol $\widetilde{M}$ has zeros, then solving the difference equation (1) becomes considerably more complicated. In general, decaying fundamental solutions do not exist. However, this can be compensated by appropriate assumptions on the growth of the data.

(39)Theorem. *Let M be any (translate of a) box spline. Then, whether or not $\widetilde{M}$ has zeros, there exists a fundamental solution λ of (1) with*

(40) $$|\lambda(j)|=O(|j|^\ell),\ |j|\to\infty,$$

for some $\ell\geq 0$ which depends on M.

The proof concerns the construction of a solution

$$A=\sum_{j\in\mathbb{Z}^s}a(j)\exp(-ij\cdot)$$

to the equation (8) and uses the existence of a tempered fundamental solution for linear constant coefficient differential operators. The main step in the proof is to show that, for an analytic P, the map

$$A\mapsto PA$$

is bounded below in the topology of the space of rapidly decreasing test functions. The corresponding estimate also holds for periodic functions P, in particular for $P=\widetilde{M}$, the symbol of (1). For this case, it can be formulated as follows, using the standard norm

$$|A|_r:=\sup_j(1+|j|)^r|a(j)|$$

on $C^\infty_{\text{periodic}}(\mathbb{R}^s)$:= the space of infinitely differentiable 2π-periodic functions of the form $A=\sum_j\exp(-ij\cdot)a(j)$.

(41)Result. *For any r, there exists an r′ (necessarily ≥ r) and a constant c so that*

$$|A|_r \le c|\widetilde{M}A|_{r'}, \qquad \forall A \in C^\infty_{\text{periodic}}. \tag{42}$$

With the aid of this Result, we now give a

Proof of (39)Theorem. By (41)Result, the functional

$$\Lambda_0 : \widetilde{M}A \mapsto a(0)$$

is well defined on $V := \{\widetilde{M}A : A \in C^\infty_{\text{periodic}}\}$ and, for some ℓ,

$$|\Lambda_0 \varphi| \le c|\varphi|_\ell, \quad \forall \varphi \in V.$$

By the Hahn-Banach Theorem, we can think of Λ_0 as the restriction to V of a bounded linear functional Λ on the space of all functions φ with $|\varphi|_\ell < \infty$. Therefore, with

$$\lambda(j) := \Lambda \exp(ij\,\cdot), \tag{43}$$

we have $|\lambda(j)| \le \|\Lambda\|\,|\exp(ij\,\cdot)|_\ell = O((1+|j|)^\ell)$. By the additivity of Λ, we have for any finitely supported a,

$$a(0) = \Lambda(\widetilde{M}A) = \sum_j \lambda(-j) \sum_k M(j-k)a(k).$$

Hence, with $a = \delta(\cdot - \ell)$,

$$\delta(-\ell) = \sum_j \lambda(-j)M(j-\ell) = \sum_j M(-\ell-j)\lambda(j),$$

or, $\delta = M_{|} * \lambda$. □

(44)Example. It is not necessary that the fundamental solution λ grow. If $|1/\widetilde{M}|$ is integrable on $[-\pi\,..\,\pi]^s$, then Λ defined by

$$\Lambda G := \frac{1}{(2\pi)^s} \int_{[-\pi..\pi]^s} \frac{G(y)}{\widetilde{M}(y)}\, dy$$

extends Λ_0, and, as expected, λ as given by (43) is the sequence of coefficients of the conjugate Fourier series for $1/\widetilde{M}$. In this case, $|\lambda(j)| = o(1)$ by the Riemann-Lebesgue Lemma. □

If a fundamental solution, λ, of the difference equation (1) has power growth as described in (40), then the corresponding fundamental function $L = M * \lambda$ satisfies the same growth estimate:

$$|L(x)| = O(|x|^\ell),\ |x| \to \infty. \tag{45}$$

Therefore, a cardinal interpolant of the form (11) exists provided that the data f decay sufficiently fast. Specifically,

(46)Corollary. *If the fundamental spline L satisfies (45), then, for any f with*
$$|f(j)| = O((1+|j|)^{-\ell'}) \quad \text{for some } \ell' > \ell + s,$$
the cardinal spline interpolant $\mathcal{L}f$, given in (11), is well-defined and satisfies
$$|\mathcal{L}f(x)| = O(|x|^\ell), \ |x| \to \infty.$$

In general, little can be said about how to define the functional Λ or how to determine ℓ for given $\widetilde{M}$. However, in some special cases more precise statements are possible. The hypothesis in the next proposition is chosen to include the ZP element (cf. (49)Example) and, more generally, the four-direction box splines with equal even multiplicities, in which case $\widetilde{M}$ has a single isolated zero.

(47)Proposition. *If $|\widetilde{M}(\cdot + z)| \ge \text{const}|\cdot|^\ell$ on $[-\pi\,..\,\pi]^s$ for some $\ell \ge s$, and $D^\beta \widetilde{M}(z) = 0$ for $|\beta| \le \ell - s$, then the functional Λ may be given by*
$$\Lambda\varphi := \frac{1}{(2\pi)^s}\int_{[-\pi..\pi]^s} \frac{\varphi(y) - \mathcal{T}_{\ell-s}\varphi(y)}{\widetilde{M}(y)}\,dy,$$
where $\mathcal{T}_k f$ is the Taylor polynomial of degree $\le k$ at z for f. The corresponding fundamental solution $\lambda : j \mapsto \Lambda(\exp(ij\cdot))$ satisfies

$$|\lambda(j)| = O\Big((1+|j|)^{\ell-s}\log(1+|j|)\Big). \tag{48}$$

Proof. We assume without loss of generality that $z = 0$. Since the Taylor polynomial $\mathcal{T}_{\ell-s}\widetilde{M}$ of $\widetilde{M}$ must vanish identically, so must $\mathcal{T}_{\ell-s}(\widetilde{M}A)$, hence the given functional Λ fulfills the requirement that $\Lambda\,\widetilde{M}A = a(0)$ for $A \in C^\infty_{\text{periodic}}(\mathbb{R}^s)$.

For the estimate on the growth of λ, the numerator of the integrand is estimated both directly and by means of the remainder in Taylor's formula:
$$\varphi(y) - \mathcal{T}_{\ell-s}\varphi(y) = \varphi(y) - \sum_{|\beta|\le\ell-s} \frac{D^\beta\varphi(0)}{\beta!}y^\beta = \sum_{|\beta|=\ell-s+1} \frac{D^\beta\varphi(\theta y)}{\beta!}y^\beta.$$
Therefore, with $\varphi = \exp(ij\cdot)$, $j \ne 0$, we obtain the bound
$$|\exp(ijy) - \mathcal{T}_{\ell-s}(\exp(ij\cdot))(y)| \le \text{const}\min\big\{(|j|\,|y|)^{\ell-s}, (|j|\,|y|)^{\ell-s+1}\big\}.$$

Using this estimate and the lower bound for $|\widetilde{M}(y)|$, the resulting integral is most easily estimated in spherical coordinates. For $|j| \ne 0$, this gives
$$|\lambda(j)| \le \text{const}\int_0^{\sqrt{s}\pi} \min\Big((|j|\varrho)^{\ell-s}, (|j|\varrho)^{\ell-s+1}\Big)\varrho^{s-1-\ell}\,d\varrho$$
$$\le \text{const}\,|j|^{\ell-s}\Big(\int_0^2 d\varrho + \int_2^{|j|\sqrt{s}\pi} d\varrho/\varrho\Big).$$

The last integration yields the estimate (48). □

(49)Example: the centered ZP element. Consider cardinal interpolation with the centered ZP element. By (III.31), the difference equation for the coefficients λ of the fundamental spline is

$$\frac{1}{2}\lambda(j)+\frac{1}{8}\big(\lambda(j-\mathbf{i}_1)+\lambda(j+\mathbf{i}_1)+\lambda(j-\mathbf{i}_2)+\lambda(j+\mathbf{i}_2)\big)=\delta(j)$$

with the corresponding symbol

$$\widetilde{M}(y)=(2+\cos(y(1))+\cos(y(2)))/4=\left(\cos(y(1)/2)^2+\cos(y(2)/2)^2\right)/2.$$

Since $\widetilde{M}$ has only an isolated zero, at $z=(\pi,\pi)$, and $\widetilde{M}$ satisfies (47)Proposition with $\ell=2=s$, the functional Λ takes the form

$$\Lambda\varphi:=\frac{4}{(2\pi)^2}\int_{[-\pi..\pi]^2}\frac{\varphi(y)-\varphi(\pi,\pi)}{2+\cos(y(1))+\cos(y(2))}dy.$$

For the particular $\varphi: x\mapsto \exp(ijx)$, we find that

$$\lambda(j)=\frac{4}{(2\pi)^2}\int_{[-\pi..\pi]^2}\frac{\exp(ijy)-(-1)^{|j|}}{2+\cos(y(1))+\cos(y(2))}dy.$$

In particular, $\operatorname{sgn}\lambda(j)=(-1)^{|j|+1}$. When $j=(\mu,\mu)$, this integral can be computed explicitly (from the formula for the discrete Green's function for the Laplacian),

$$(50)\qquad \lambda(\mu,\mu)=\frac{-8}{\pi}\Big(1+\frac{1}{3}+\frac{1}{5}+\frac{1}{7}+\cdots+\frac{1}{2|\mu|-1}\Big),\qquad \mu\neq 0.$$

In particular, the bound $|\lambda(j)|=O(\log|j|)$ in (48) is attained for the centered ZP element.

Level curves, $\widetilde{M}(y)=c=0{:}.05{:}1$, for the symbols of the centered ZP element and the centered four-direction box spline $M^c_{2,0,3,1}$ are given in (51)Figure (a) and (b) respectively. It would be ideal if the only zeros of the symbols for the centered four-direction box splines were the zeros at (π,π) dictated by the linear dependence (cf. (30)), for then (47)Proposition might tell the whole story. This hope is quickly dismissed since the symbol for $M^c_{2,0,3,1}$ is negative in the shaded region. □

Bivariate four-direction box splines with even multiplicities. Even though the fundamental solution λ of the difference equation may grow, it turns out that the estimate (45) for the growth of the fundamental spline may be overly pessimistic. Such is the case for the bivariate four-direction box splines with equal even multiplicities.

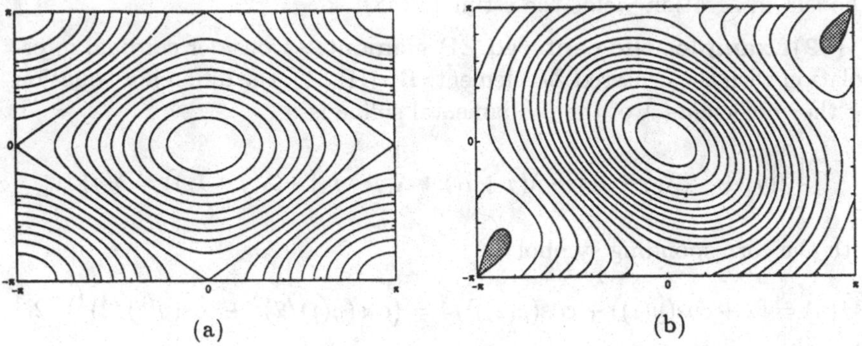

(51)Figure. Level curves for $\widetilde{M}^c_{1,1,1,1}$ and $\widetilde{M}^c_{2,0,3,1}$.

(52)Proposition. *For a bivariate centered four-direction box spline with equal even multiplicities, there is a fundamental solution of (1) for which the corresponding fundamental function L vanishes at infinity; i.e.,*

$$|L(x)| = o(1), \quad |x| \to \infty.$$

Proof. Let $M = M^c_\Xi$, where Ξ consists of the four directions $\xi_1 = \mathbf{i}_1$, $\xi_2 = \mathbf{i}_2$, $\xi_3 = \mathbf{i}_1 + \mathbf{i}_2$, and $\xi_4 = \mathbf{i}_1 - \mathbf{i}_2$, each with multiplicity ℓ. Since ℓ is even, we have

$$\widehat{M}(2\pi v) = \prod_{\mu=1}^{4} \left(\frac{\sin(\pi v\xi_\mu)}{\pi v\xi_\mu}\right)^\ell \geq 0$$

with equality if and only if $2\pi v = (\pi, \pi) \bmod 2\pi\mathbb{Z}^2$.

We wish to show that the hypothesis of (47)Proposition applies to the symbol $\widetilde{M}$. For this we need good estimates on the symbol in a neighborhood of its zero, $v = (1/2, 1/2) \bmod \mathbb{Z}^2$. As the transformations $\mathbf{i}_1 \mapsto -\mathbf{i}_1$, $\mathbf{i}_2 \mapsto -\mathbf{i}_2$, and $\mathbf{i}_1 \mapsto \mathbf{i}_2$ each leave Ξ unchanged up to signs of columns, the corresponding symmetries of $\widetilde{M}$ implied by (34) show that it suffices to restrict v to the square $[0\,..\,1/2]^2$.

The symbol $\widetilde{M}_x$, for the centered box spline translated by x, will enter naturally later in the proof. The required estimates for $\widetilde{M} = \widetilde{M}_0$ and $\widetilde{M}_x$ are obtained in the same way using the Poisson summation formula

$$\begin{aligned} \widetilde{M}_x(2\pi v) &= \sum_{j\in\mathbb{Z}^2} M(j - x)\exp(-2\pi i j v) \\ &= \sum_{j\in\mathbb{Z}^2} \widehat{M}(2\pi(j + v))\exp(-2\pi i x(v + j)). \end{aligned} \tag{53}$$

For $v \in [0\,..\,1/2]^2$, we have $0 \le v\xi_1, v\xi_2 \le 1/2$, $0 \le v\xi_3 \le 1$, and $-1/2 \le v\xi_4 \le 1/2$, while, for any $w \in [-1\,..\,1]$,

$$\left|\frac{\sin(\pi w)}{\pi w}\right| \asymp \min\left\{1, \frac{1-|w|}{|w|}\right\},$$

where, as usual, $a \asymp b$ stands for the inequalities $C_1 a \le b \le C_2 a$, with some positive constants C_1, C_2. Therefore,

$$\widehat{M}(2\pi v) \asymp (1 - v\xi_3)^\ell \asymp |(\pi,\pi) - 2\pi v|^\ell, \qquad \forall v \in [0\,..\,1/2]^2, \tag{54}$$

where the last relation follows because the distance d_1 of a point $v \in [0\,..\,1/2]^2$ to the line $1 - v\xi_3 = 0$ and the distance d_2 of this point to $(1/2, 1/2)$ satisfy the relation $d_1 = d_2 \cos\theta$ for some $\theta \in [0\,..\,\pi/4]$ (see (55)Figure).

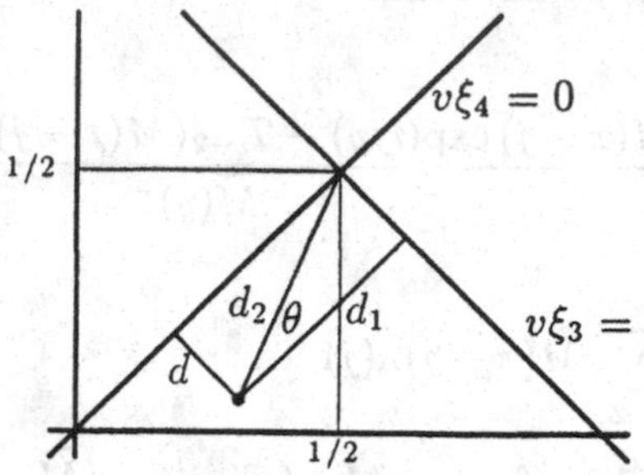

(55)Figure. The distance of a point in $[0\,..\,1/2]^2$ to the lines $1 - v\xi_3 = 0$ and $v\xi_4 = 0$ and to their intersection.

For $j \ne 0$ and $v \ne (1/2, 1/2)$,

$$\frac{\widehat{M}(2\pi(j+v))}{\widehat{M}(2\pi v)} = \left(\frac{v\xi_1}{(j+v)\xi_1}\right)^\ell \left(\frac{v\xi_2}{(j+v)\xi_2}\right)^\ell \left(\frac{v\xi_3}{(j+v)\xi_3}\right)^\ell \left(\frac{v\xi_4}{(j+v)\xi_4}\right)^\ell.$$

Since for positive integers r, the functions $w \mapsto w/(r \pm w)$ are increasing in $(-1\,..\,1)$, the first two factors on the right are bounded above by

$$(2j(1) + 1)^{-\ell}(2j(2) + 1)^{-\ell},$$

while the last two factors are bounded above by 1 unless $j\xi_3 = -1$. If $j\xi_3 = -1$, then $j\xi_4 \ne 0$, $|(j+v)\xi_4| \ge 1/2$, and $|v\xi_4| \le (1 - v\xi_3)$ (the distance d to the line $v\xi_4 = 0$ is less than the distance d_1 to the line $v\xi_3 = 1$, see (55)Figure). Thus in any case, the last two factors are bounded by 2^ℓ; therefore, for $v \in [0\,..\,1/2]^2 \backslash (1/2, 1/2)$ and $j \in \mathbb{Z}^s \backslash 0$,

$$\frac{\widehat{M}(2\pi(j+v))}{\widehat{M}(2\pi v)} \le 2^\ell (2j(1) + 1)^{-\ell}(2j(2) + 1)^{-\ell}. \tag{56}$$

Consequently, for $v \in [0\,..\,1/2]^2 \backslash (1/2, 1/2)$,

$$
(57)\qquad
\begin{aligned}
|\widetilde{M}_x(2\pi v)| \le \widetilde{M}(2\pi v) &= \widehat{M}(2\pi v)\Big(1 + \sum_{j\in\mathbb{Z}^2\backslash 0} \frac{\widehat{M}(2\pi(j+v))}{\widehat{M}(2\pi v)}\Big) \\
&\asymp |(\pi,\pi) - 2\pi v|^{\ell}.
\end{aligned}
$$

It follows that, for every x, the Taylor polynomial $T_{\ell-2}\widetilde{M}_x$ for $\widetilde{M}_x$ at (π,π) vanishes identically.

In particular, the hypothesis of (47)Proposition holds for our $\widetilde{M}$ and ℓ (since $s = 2$). Thus, the functional Λ may be taken as given there, with $z = (\pi,\pi)$. This time however, we assess the behavior of

$$L = M * \lambda.$$

For $\lambda(j) = \Lambda \exp(ij\,\cdot\,)$, we have that

$$
M(x-j)\lambda(j) = \\
\frac{1}{(2\pi)^2}\int_{[-\pi..\pi]^2} \frac{M(x-j)\exp(ijy) - T_{\ell-2}(M(x-j)\exp(ij\,\cdot\,))(y)}{\widetilde{M}(y)}\,dy,
$$

and

$$
(58)\qquad
\begin{aligned}
L(x) &= \sum_{j\in\mathbb{Z}^s} M(x-j)\lambda(j) \\
&= \frac{1}{(2\pi)^2}\int_{[-\pi..\pi]^2} \frac{\widetilde{M}_{-x}(y) - T_{\ell-2}(\widetilde{M}_{-x})(y)}{\widetilde{M}(y)}\,dy \\
&= \frac{1}{(2\pi)^2}\int_{[-\pi..\pi]^2} \frac{\widetilde{M}_{-x}(y)}{\widetilde{M}(y)}\,dy.
\end{aligned}
$$

The relations in (57) show that the last integrand is bounded; i.e., $|L(x)| = O(1)$.

For the $o(1)$ result, we use (53) to write $\widetilde{M}_{-x}(2\pi v)$ as a sum, and factor out $\widehat{M}(2\pi v)$ to obtain

$$
\frac{\widetilde{M}_{-x}(2\pi v)}{\widetilde{M}(2\pi v)} = \\
\frac{\widehat{M}(2\pi v)}{\widetilde{M}(2\pi v)}\Big(\exp(-2\pi i x v) + \sum_{j\in\mathbb{Z}^2\backslash 0} \frac{\widehat{M}(2\pi(j+v))\exp(-2\pi i x(v+j))}{\widehat{M}(2\pi v)}\Big).
$$

The first quotient on the right is a bounded function by (54) and (57). Since the series is absolutely summable (cf. (56)), given $\varepsilon > 0$, there is an $N > 0$ so that

$$\sum_{|j|\ge N} |\widehat{M}(2\pi(j+v))\exp(-2\pi i x(v+j))|/\widehat{M}(2\pi v) \le \varepsilon$$

holds independently of x and v. Therefore, the last integral in (58) consists of a term corresponding to the sum with $|j| \geq N$ and finitely many terms of the form

$$\frac{1}{(2\pi)^2} \int_{[-\pi..\pi]^2} \frac{\widehat{M}(2j\pi + y)\exp(-ix(2j\pi + y))}{\widetilde{M}(y)}\,dy,$$

and these can be recognized as Fourier transforms of integrable functions (cf. (54), (56), and (57)). This first term is $O(\varepsilon)$, while the latter terms go to 0 as $|x| \to \infty$ by the Riemann-Lebesgue Lemma. □

Although the ZP element does not fit into the framework of (52)Proposition, it does have the $o(1)$ decay. This is illustrated in (59)Figure where its graph is given on $[0\,..\,6]^2$. The decay, which with some difficulty can be seen to be at least $O\big((1+\log|x|)/(1+|x|)\big)$, is a result of the fact that the coefficients of the polynomial pieces making up L are determined from the *differences* of the sequence λ rather than from λ itself.

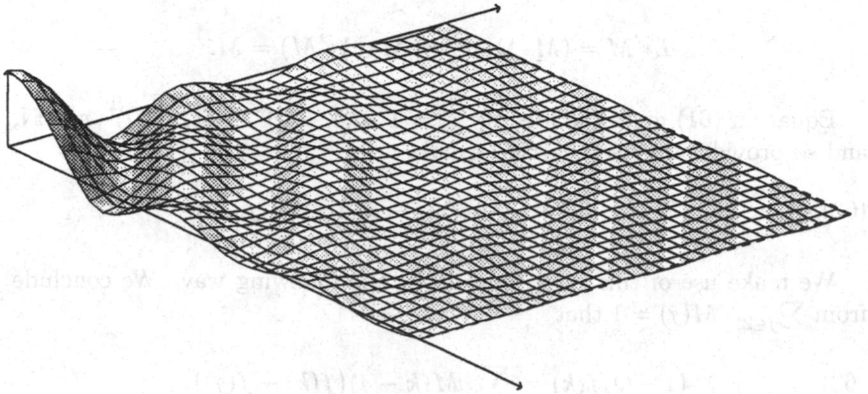

(59)Figure. Fundamental spline for $M^c_{1,1,1,1}$.

The approximation order of cardinal interpolation. With $\sigma_h f = f(\cdot/h)$, denote by

$$\mathcal{L}_h f := \sigma_h \mathcal{L} \sigma_{1/h} f = \sum_j L(\cdot/h - j) f(jh)$$

the cardinal interpolant with respect to the scaled lattice $h\mathbb{Z}^s$. We show that, for sufficiently smooth f, $\mathcal{L}_h$ has the optimal approximation order (cf. (III.1)Proposition).

(60)Theorem. *For any function $f \in C^{\ell+s+m(\Xi)+1}$ with support in the compact set Ω,*

$$\|f - \mathcal{L}_h f\|_{\infty,\Omega} = O(h^{m(\Xi)+1}),$$

where ℓ is defined in (45).

The assumption that f have compact support is convenient because of the polynomial growth of the fundamental function L. For an arbitrary smooth function f, the theorem can be applied to $f\chi$ where χ is a smooth cut-off function, and this yields convergence of cardinal interpolants on arbitrary bounded domains.

Proof. We begin by showing that, for any (univariate) polynomial p with $p(0) = 0$,

$$\mathcal{L}p(Q) = p(Q), \tag{61}$$

where $Qf := M*'f$. For this, it is sufficient to show that $\mathcal{L}Q = Q$, i.e., that $L*'M = M$. But this follows at once from the fact that, by (10) and the definition of L,

$$L*'M = (M*\lambda)*'M = M*(\lambda*'M) = M.$$

Equation (61) gives the identity $\mathcal{L} = 1 - (1-Q)^r + \mathcal{L}(1-Q)^r, r \in \mathbb{N}$, and so provides the useful error formula

$$f - \mathcal{L}f = (1-Q)^r f - \mathcal{L}(1-Q)^r f. \tag{62}$$

We make use of this error formula in the following way. We conclude from $\sum_{j\in\mathbb{Z}^s} M(j) = 1$ that

$$(1-Q)f(k) = \sum_{j\in\mathbb{Z}^s} M(k-j)\big(f(k) - f(j)\big) \tag{63}$$

is a first-order difference, hence $(1-Q)^r f(k)$ is an rth order difference of f, hence boundable in terms of $D^r f$. Thus the second term in (62) is of the order $(R_f + rR_M)^{\ell+s}\|D^r f\|$, with R_f the radius of a ball containing the support of f. As to the first term, we recall from the proof of (III.6)Proposition that $(1-Q)$ is degree reducing, i.e.,

$$Qp \in p + \Pi_{<\deg p} \quad \forall p \in \Pi_{<m(\Xi)+1}.$$

This implies that, for $r \ge m(\Xi) + 1$,

$$|(1-Q)^r f(x)| \le \text{const}_{M,r} \inf_{p\in\Pi_{<m(\Xi)+1}} \sup_{|y-x|<rR_M} |(f-p)(y)|$$
$$\le \text{const}\,(rR_M)^{m(\Xi)+1}\|D^{m(\Xi)+1} f\|.$$

It follows that

$$(64)\qquad \|f-\mathcal{L}f\| \le \text{const}\,\|D^{m(\Xi)+1}f\| + \text{const}\,(R_f+\text{const})^{\ell+s}\|D^r f\|,$$

with the various constants independent of f, and $r \ge m(\Xi)+1$.

In consequence, the error of scaled interpolation is of order $O(h^{m(\Xi)+1})+O(h^{-\ell-s}h^r)$, and this is $O(h^{m(\Xi)+1})$ if f is smooth enough; for example, if f has $\ell+s+m(\Xi)+1$ continuous derivatives. □

Notes. The theory of cardinal interpolation by box splines was started (in [de Boor, Höllig, Riemenschneider'83,'85b]) in order to lift the beautiful theory that I. J. Schoenberg had developed for univariate cardinal splines (see [Schoenberg'73a]) to several variables, in a nontensor product setting. Subsequent developments have been surveyed in [Jetter'87a], [Chui'88], [Riemenschneider'89], and [Jetter'92a].

As to specific results, (12)Theorem is standard; it can be found in, e.g., [de Boor, Höllig, Riemenschneider'85b]) together with a rather complete discussion of the bivariate three-direction box spline. In particular, it is shown there that interpolation is correct with the centered three-direction box spline for any T_r as long as at least two of the multiplicities are nonzero. The case when all multiplicities are odd is discussed in (35)Example to give the flavor of the arguments involved.

For translates of three-direction box splines, the simple geometric argument as used in (26)Example must be replaced by more complicated arguments, but all results point to the region J (see (27)Figure(b)) as constituting the collection of all the translation vectors (mod $\mathbb{Z}^2$) that lead to correct interpolation regardless of multiplicity. Papers that deal with this question are [Sivakumar'90a], [Stöckler'89a], [Chui, Stöckler, Ward'89a,'89b, '91a], and [Binev, Jetter'91,'92a].

The form of (28)Proposition is basically due to [Ron'89c] and, as noted in the proof, the proposition is also obtainable from the corresponding result in [Dahmen, Micchelli'83b] by an application of the Poisson summation formula. Ron proved the theorem for compactly supported distributions, something that we shall make use of in Chapter VI:

(65)Theorem. *The shifts of a compactly supported distribution φ are linearly independent if and only if there is no $y \in \mathbb{C}^s$ for which $\widehat{\varphi}(y+2\pi j)=0$ for all $j \in \mathbb{Z}^s$.*

Notice that this general theorem requires $y \in \mathbb{C}^s$ whereas (28)Proposition requires only $y \in \mathbb{R}^s$. This may be explained by pointing out that the real part of any complex zero of the Fourier transform of a box spline is also a zero. Incidentally, [Ron'89c] also shows that existence of a real zero of the

Fourier transform of φ is equivalent to the existence of a nontrivial element of $\ker \varphi^*$ which grows at most polynomially.

The use of Poisson's summation formula, in the proof of (28)Proposition, and of (31), is justified by the inequality

$$(66)\qquad |\widehat{M}(y)| \le \prod_{\xi\in\Xi} \frac{2}{1+|y\xi|} \le \frac{2^{\#\Xi}}{(1+\text{const}\,|y|)^{m(\Xi)}} \sum_{\mathrm{B}\in\mathbb{B}(\Xi)} \prod_{\xi\in\mathrm{B}} \frac{1}{1+|y\xi|},$$

since this shows that $\widehat{M}$ is absolutely integrable on $\mathbb{R}^s$ as soon as M is continuous, i.e., as soon as $m(\Xi) > 0$ (since it shows it to be bounded by a finite sum of functions, each of which decaying like $O(\prod_{\xi\in\mathrm{B}} 1/(1+y\xi)^{1+1/s})$ for some basis $\mathrm{B} \in \mathbb{B}(\Xi)$).

For the proof of (66), we first observe that, for any $y \ne 0$, the set $\{\xi \in \Xi : y\xi = 0\}$ is contained in some hyperplane $H = H_y \in \mathbb{H}(\Xi)$. Therefore, by the compactness of the unit ball,

$$\min_{|y|=1} \max_{H\in\mathbb{H}(\Xi)} \min_{\xi\in\Xi\backslash H} |y\xi| \ge \text{const} > 0$$

for some positive const. Hence, for any $y \ne 0$ there is a hyperplane $H = H_y \in \mathbb{H}(\Xi)$ with

$$\text{const}\,|y| \le |y\xi|, \qquad \forall \xi \in \Xi\backslash H.$$

Finally, for $y \ne 0$ and its associated H, select $\zeta = \zeta_y \in \Xi\backslash H$ and some basis $\mathrm{B} = \mathrm{B}_y$ from $\zeta \cup (\Xi \cap H)$. Since $\#(\Xi\backslash H) \ge m(\Xi) + 1$ regardless of $H \in \mathbb{H}(\Xi)$, the inequality (66) holds for y already with the only summand on the right hand side the one corresponding to this B.

Since the symbol for a nontrivially bivariate centered three-direction box spline was shown in [de Boor, Höllig, Riemenschneider'85b] to be positive, the assumption of nonnegativity of the symbol can be dropped from (32)Proposition in case $s = 2$. This led to the conjecture that for centered box splines, linear independence of the shifts is also sufficient for correctness of cardinal interpolation. However, with the aid of the fundamental domain described in Chapter V, [de Boor, Höllig, Riemenschneider'85a] found a box spline having linearly independent shifts for which cardinal interpolation is not correct.

There have been several attempts to deal with *singular cardinal interpolation*, mainly focused on the goal of accommodating the bivariate four-direction box spline. The approach of [Jetter, Riemenschneider'86a,'87] was to consider the interpolation on a sublattice of the integer lattice $\mathbb{Z}^s$ with interpolation from the space generated by the shifts of the box spline over the same sublattice. They showed that once cardinal interpolation was correct for some sublattice, then it was also correct for interpolation formulated with any coarser sublattice. Since approximation order can only get worse when one goes to a coarser sublattice and interpolation will be correct for any sufficiently coarse lattice (i.e., as soon as the supports of the corresponding shifts are disjoint), the object would be to find the "finest" sublattice on which interpolation is correct. For the bivariate four-direction box spline, this finest sublattice is the **even lattice**, $Q\mathbb{Z}^2 = \{(k,\ell) : k+\ell \in 2\mathbb{Z}\}$, generated by the matrix $Q = \left[\begin{smallmatrix} 1 & 1 \\ 1 & -1 \end{smallmatrix}\right]$. Jetter & Riemenschneider show that cardinal interpolation on the even lattice is correct for any nontrivially bivariate four-direction box spline. From the equation (I.23) relating the box splines for Ξ and $Q\Xi$, it follows that cardinal interpolation on the sublattice $Q\mathbb{Z}^s$ with the box spline M_Ξ is equivalent to cardinal interpolation on $\mathbb{Z}^s$ with the box spline $M_{Q^{-1}\Xi}$. The latter observation sparked the interest in box splines based on rational directions; see [Jia, Sivakumar'90], [Ron, Sivakumar'93], [Sivakumar'91a], [Ron'92a].

A more direct approach to singular interpolation is to resolve the singularity of $1/\widetilde{M}$. This was first accomplished for some particular types of singularities (that included the singularity in the case of the ZP element) in [Chui, Diamond, Raphael'88b] through a careful analysis of the Neumann series associated with the symbol. The distributional approach presented here for (39)Theorem and (46)Corollary was used in [de Boor, Höllig, Riemenschneider'89]. It relies on (41)Result, i.e., on the solution, by [Łojasiewicz'59] and [Hörmander'58], of the division problem of L. Schwartz, i.e., the construction of a tempered fundamental solution for constant-coefficient differential operators. The basic method for resolving the isolated singularities in (47)Proposition is standard. It is carried out in full detail in [de Boor, Höllig, Riemenschneider'89] for the symbol of the bivariate discrete Laplacian (which relates quite easily to the symbol for the ZP element). The formula (50) for the Green's function for the discrete Laplacian can already be found in [McCrea, Whipple'40], as has been pointed out to us by R.E. Lynch.

The decay of a fundamental solution for cardinal interpolation by certain four-direction box splines as exhibited in (52)Proposition was also found by [Jetter, Stöckler'91a]. They showed that, for multiplicities r with $r(1) = r(2)$ and $r(3) = r(4)$, the symbol has only the isolated zero at (π,π), and that $\widehat{M}/\widetilde{M}$ belongs to $L_1(\mathbb{R}^2)$, from which a fundamental solution is derived by taking the inverse Fourier transform.

The material on *approximation order of cardinal interpolation* is taken from [de Boor, Höllig, Riemenschneider'89]. Both the papers [Chui, Diamond, Raphael'88b] and [de Boor, Höllig, Riemenschneider'89] showed the optimality of the approximation order for approximation by the scaled cardinal interpolation operators considered there. [Chui, Jetter, Ward'87] established the optimal approximation order in $\mathbf{L}_2$ when interpolation is correct.

only get worse when one goes to a coarser sublattice and that prediction will be correct for any sufficiently coarse sublattice if [illegible] as soon as the support of the corresponding [illegible]. [illegible] would also find the finest sublattice on which interpolation is correct. For the three- or four-direction box spline, this finest sublattice is the even lattice, $\mathbb{Z}^2 = \{(i,j) : i + j \in 2\mathbb{Z}\}$, as covered by the matrix [illegible]. Further, one can show that cardinal interpolation on the even lattice is correct for any nondegenerate four-direction box spline [illegible] equation (1.8) [illegible] the boundaries for [illegible]. It follows that cardinal interpolation on the sublattice $\Xi\mathbb{Z}^2$ with the box spline M_Ξ is equivalent to cardinal interpolation [illegible] with the box spline [illegible]. The [illegible] of [illegible] interpolation [illegible] has been [illegible] studied [illegible] [illegible] [illegible] [illegible].

[illegible] approach to cardinal interpolation is to resolve the singularity of [illegible]. This was first accomplished for one particular type of singularities [illegible] in the case of the [illegible] in [illegible] Riemenschneider [illegible] a careful analysis of the [illegible] associated with the symbol. The [illegible] approach [illegible] and [illegible] was used in [illegible] [illegible] in [illegible] Riemenschneider [illegible] in [illegible] [illegible] and [illegible] the [illegible] of [illegible] the [illegible] fundamental solution [illegible] difference operators. The [illegible] of [illegible] in [illegible] in full detail [illegible] the [illegible] of [illegible] the symbol of [illegible] relates quite [illegible] for the [illegible]. The [illegible] for the [illegible] for the [illegible] as has been [illegible] by R. [illegible].

The [illegible] fundamental solution for cardinal interpolation by [illegible] four-direction box splines as [illegible] in (3. [illegible] was also found by [illegible]. However, [illegible] with [illegible] and [illegible] the [illegible] only the [illegible] [illegible] [illegible] fundamental solution [illegible] linear [illegible].

The [illegible] approximation order of cardinal interpolation [illegible] [illegible] Höllig, Riemenschneider [illegible] and [illegible] [illegible] the optimality of the approximation order [illegible] cardinal interpolation [illegible] established [illegible] approximation order [illegible] when interpolation [illegible] correct.

V

Approximation by cardinal splines & wavelets

In this chapter, we study the approximation by cardinal spline series. We look at just two possibilities. In the first part, we consider approximation as the degree tends to infinity. In the second part, we keep the degree (i.e., the direction matrix used) fixed, but allow the mesh width to go to zero, and this includes a discussion of box spline wavelets.

Fundamental domains. A striking result from the univariate spline theory states that a function f in $\mathbf{L}_2(\mathbb{R})$ is the norm limit of a sequence of cardinal splines g_n of degree $\le n$ as n tends to infinity if and only if f is an entire function of exponential type $\le \pi$. Moreover, the cardinal spline interpolants $\mathcal{L}_n f$ to f form such a sequence g_n. For multivariate box splines the role of the degree is played by $\#\Xi$ and, in generalizing the univariate result, one is faced with the problem of how to choose the sequence of matrices Ξ with $\#\Xi \to \infty$. A simple choice is to consider, for a fixed matrix Ξ, the sequence Ξ_r, $r \in \mathbb{N}$, consisting of r copies of Ξ (cf. (II.40)). We denote the box splines corresponding to Ξ_r by M_{*_r}; i.e.,

$$(1) \qquad M_{*_r} := M_{\Xi_r} = \underbrace{M_\Xi * \ldots * M_\Xi}_{r \text{ factors}}.$$

The underlying reason for the univariate result mentioned above is that, for the centered univariate cardinal B-spline M_n^c of order n,

$$\frac{\widehat{M_n^c}}{\widetilde{M_n^c}} \to \chi_{(-\pi..\pi)} \ a.e., \qquad \text{as } n \to \infty.$$

For the box spline $M = M_\Xi$, the role of the interval $(-\pi \,..\, \pi)$ is played by the set

$$\text{(2)} \qquad \Omega_M := \{y \in \mathbb{R}^s : |\widehat{M}(y + 2\pi j)| < |\widehat{M}(y)|, \ \forall j \in \mathbb{Z}^s \backslash 0\}.$$

Since $\widehat{M}_{*_r} = \left(\widehat{M_\Xi}\right)^r$, this set depends only on Ξ, and not on r; moreover, it is the same set if M is replaced by any translate $M(\cdot - x_0)$. Therefore, the notation Ω_Ξ will also be used to denote the set Ω_M for any box spline M based on the matrices Ξ_r, and we will often drop either subscript if the function or matrix is understood.

The importance of Ω_M stems from the fact that it is a **fundamental domain** in the sense of Fourier analysis, i.e., $\Omega = \Omega_M$ satisfies the three conditions

$$\text{(3)} \qquad \begin{aligned} &(a) \quad \text{meas}\left(\mathbb{R}^s \backslash \cup_{j \in \mathbb{Z}^s} (\Omega + 2\pi j)\right) = 0; \\ &(b) \quad \Omega^- \cap (\Omega + 2\pi j) = \emptyset, \qquad j \in \mathbb{Z}^s \backslash 0; \\ &(c) \quad \text{vol}(\Omega) = (2\pi)^s. \end{aligned}$$

Here, and throughout this chapter, we denote by A^- the closure of a subset A of a metric space.

(4)Proposition. *If M is a nontrivial compactly supported* $\mathbf{L}_2$*-function, then $\Omega = \Omega_M$ is an open fundamental domain.*

Proof. Since M is compactly supported, it also belongs to $\mathbf{L}_1(\mathbb{R}^s)$. Therefore, $\widehat{M}$ is uniformly continuous and

$$\text{(5)} \qquad |\widehat{M}(y)| = o(1), \qquad \text{as } |y| \to \infty,$$

by the Riemann-Lebesgue lemma. This implies that, for any $y \in \mathbb{R}^s$, the function

$$\mathbb{Z}^s \to \mathbb{R} : j \mapsto |\widehat{M}(y + 2\pi j)|^2$$

has a maximum, hence Ω_M is open.

Further, if this maximum is unique, say at $j = j_*$, then $y \in \Omega - 2\pi j_*$. Thus the complement of $\cup_{j \in \mathbb{Z}^s}(\Omega + 2\pi j)$ is contained in sets of the form

$$\{y : f_j(y) = f_k(y), \text{ some } j \neq k\},$$

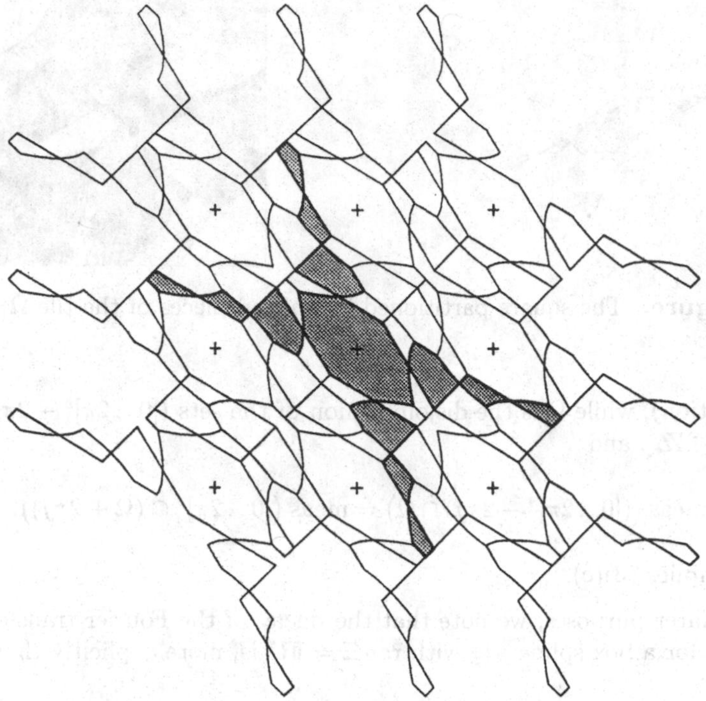

(6)Figure. The tiling provided by Ω_Ξ with $\Xi = \begin{bmatrix} 1 & 3 \\ 3 & 1 \end{bmatrix}$.

where $f_j := |\widehat{M}(\cdot + 2\pi j)|^2$. But, by the Paley-Wiener theorem, the Laplace-Fourier transform of the compactly supported function M is the restriction to $\mathbb{R}^s$ of an entire function on $\mathbb{C}^s$, and therefore each f_j is a real analytic function on $\mathbb{R}^s$. This shows that the complement of $\cup_{j \in \mathbb{Z}^s}(\Omega + 2\pi j)$ is contained in the union of the zero sets of the real analytic functions $f_j - f_k$, $j \neq k$. Since there are only countably many such functions, and the zero set of a nontrivial real analytic function is of measure zero, the verification of (3)(a) is complete once we show that $f_j - f_k$ cannot vanish identically. This follows since $f_j - f_k = 0$ would imply that $|\widehat{M}|^2$ is periodic in the direction $j - k$, and this would contradict (5).

To verify (3)(b), let $y = \lim_{\nu \to \infty} y_\nu$ with $y_\nu \in \Omega$. Then the assumption that $y - 2\pi j$ belongs to Ω for some $j \neq 0$ leads to the contradiction

$$\begin{aligned} 1 &> |\widehat{M}((y - 2\pi j) + 2\pi j)/\widehat{M}(y - 2\pi j)| \\ &= \lim_{\nu \to \infty} |\widehat{M}(y_\nu)/\widehat{M}(y_\nu - 2\pi j)| \geq 1. \end{aligned}$$

We conclude that, up to a set of measure zero, $[0\,..\,2\pi]^s$ is the disjoint union of the sets $[0\,..\,2\pi]^s \cap (\Omega + 2\pi j)$ with $j \in \mathbb{Z}^s$ (see (7)Figure for an

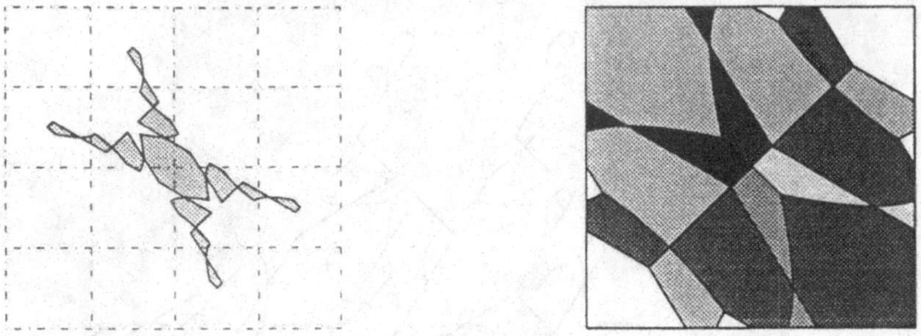

(7)Figure. The square partitioned by shifted pieces of the tile $\Omega_{\left[\begin{smallmatrix}1&3\\3&1\end{smallmatrix}\right]}$.

illustration), while Ω is the disjoint union of the sets $([0\,..\,2\pi]^s - 2\pi j) \cap \Omega$ with $j \in \mathbb{Z}^s$, and

$$\text{meas}\,\big(([0\,..\,2\pi]^s - 2\pi j) \cap \Omega\big) = \text{meas}\,\big([0\,..\,2\pi]^s \cap (\Omega + 2\pi j)\big).$$

This implies (3)(c). □

For later purposes, we note that the decay of the Fourier transform at infinity for a box spline M_Ξ with $\text{ran}\,\Xi = \mathbb{R}^s$ is, more explicitly than (5),

$$(8) \qquad |\widehat{M_\Xi}(y)| = O\Big(\frac{1}{|y|}\Big).$$

This follows from the formula (I.17) for the Fourier transform since $|(1 - \exp(-i\xi y))/i\xi y)| \le 1$ and

$$(9) \qquad \max_{\xi\in\Xi} |\xi y| \ge C|y| \qquad \text{with} \quad C := \max_{Z\in\mathcal{B}(\Xi)} \min_{|y|=1} \max_{\zeta\in Z} |y\zeta| > 0.$$

Properties of Ω_Ξ. The main additional properties of Ω are summarized in

(10)Proposition. *If $M = M_\Xi$ for some $\Xi \in \mathbb{Z}^{s\times n}$ with $\text{ran}\,\Xi = \mathbb{R}^s$, then Ω is bounded, symmetric with respect to the origin, and does not intersect any of the hyperplanes*

$$(11) \qquad \{y \in \mathbb{R}^s : y\xi = 2\pi\nu\}, \qquad \xi \in \Xi, \quad \nu \in \mathbb{Z}\backslash 0.$$

Moreover, if $Q \in \mathbb{Z}^{s\times s}$ is a unimodular matrix which leaves $\Xi \cup -\Xi$ invariant, then

$$(12) \qquad \Omega = Q^T\Omega.$$

Proof. Since $\Xi \in \mathbb{Z}^{s\times n}$, $\widehat{M}$ vanishes on the hyperplanes (11) and therefore these hyperplanes do not intersect Ω. In particular, $\widehat{M}$ vanishes at the points $2\pi j$, $j \in \mathbb{Z}^s\backslash 0$, and this together with $\widehat{M}(0) = 1$ implies that Ω contains the origin. The symmetries of Ω come from the action of unimodular integer matrices on Ξ; the formula (I.23): $\widehat{M}_{Q\Xi} = \widehat{M}_\Xi \circ Q^T$ implies that

$$\Omega_\Xi = Q^T \Omega_{Q\Xi} \quad \text{for unimodular} \quad Q \in \mathbb{Z}^{s\times s}. \tag{13}$$

This formula with $Q = -\mathbb{1}$ and (I.19) show that $|\widehat{M}(y)| = |\widehat{M}(-y)|$, and this provides the symmetry through the origin. More generally, (13) and (I.19) imply (12).

For the boundedness of Ω, it suffices to show that, for any $y \in \Omega$, there exists $\mathrm{B} \in \mathcal{B}(\Xi)$ such that

$$|y\xi| \leq \text{const}, \quad \forall \xi \in \mathrm{B},$$

with const independent of y and B. To this end, given $y \in \Omega\backslash 0$, choose $s-1$ columns $\mathrm{Z} \subset \Xi$ of full rank for which $\|\mathrm{Z}^T y\|_1$ is minimal, let H be the hyperplane spanned by Z, and choose $\eta \in \Xi\backslash H$ so that $|y\eta|$ is minimal. Since there are only finitely many bases in $\mathcal{B}(\Xi)$, the set of all possible coefficients in representations of elements $\xi \in \Xi$ by bases $\mathrm{B} \in \mathcal{B}(\Xi)$ is a bounded set. In particular, for $\mathrm{B} = \eta \cup \mathrm{Z}$, every $\xi \in \Xi\backslash(H \cup \eta)$ has a representation

$$\xi = \xi_{\mathrm{z}} + a(\xi)\eta, \quad \xi_{\mathrm{z}} \in H,$$

with coefficients $a(\xi) = a(\xi, \eta, \mathrm{B})$ from this bounded set.

By the minimality of Z, we must have

$$|y\eta| \geq |y\xi|, \quad \forall \xi \in \mathrm{Z}.$$

Therefore, it only remains to bound $|y\eta|$ independently of the choices of H and η. For this, we may assume that $|y\eta| \geq 4\pi|\det \mathrm{B}|$. The bound is now obtained by appealing to the definition of Ω with j chosen as the solution to the system

$$\begin{aligned} j\xi &= 0, \qquad \xi \in \mathrm{Z}, \\ j\eta &= -t, \end{aligned}$$

where $\operatorname{sgn}(y\eta)t$ is the largest multiple of $|\det \mathrm{B}|$ less than $|y\eta|/2\pi$. Then $j \in \mathbb{Z}^s\backslash 0$, $j\xi = 0$ for all $\xi \in H$, $|y\eta + 2\pi j\eta| \leq 2\pi|\det \mathrm{B}|$, and $2\pi|t| \leq |y\eta|$.

Thus, from the definition of Ω and our choices of H, η and j, we have

$$\begin{aligned}
1 \ge \left|\frac{\widehat{M}(y+2\pi j)}{\widehat{M}(y)}\right| &= \frac{|y\eta|}{|y\eta+2\pi j\eta|} \prod_{\xi\in\Xi\backslash(H\cup\eta)} \frac{|y\xi|}{|y\xi+2\pi j\xi|}\\
&\ge \frac{|y\eta|}{2\pi|\det \mathrm{B}|} \prod_{\xi\in\Xi\backslash(H\cup\eta)} \frac{|y\xi|}{|y\xi|+2\pi|a(\xi)|\,|j\eta|}\\
&\ge \frac{|y\eta|}{2\pi \operatorname{vol}(\Xi\mathsf{L})} \prod_{\xi\in\Xi\backslash(H\cup\eta)} \frac{|y\eta|}{|y\eta|+2\pi|a(\xi)|\,|t|}\\
&\ge \frac{|y\eta|}{2\pi \operatorname{vol}(\Xi\mathsf{L})} \prod_{\xi\in\Xi\backslash(H\cup\eta)} \frac{1}{1+|a(\xi)|} \ge \operatorname{const}|y\eta|,
\end{aligned}$$

for some absolute (positive) constant. □

When the shifts of the box spline are linearly independent, one additional property of Ω appears:

(14)Proposition. *$\widehat{M}$ has no zeros in Ω^- if and only if the shifts of M are linearly independent; i.e., if and only if Ξ is a unimodular matrix.*

Proof. Recall that by (IV.28)Proposition, the shifts of M are linearly independent if and only if for all $y \in [0\,..\,2\pi)^s$ there is some $j \in \mathbb{Z}^s$ for which $\widehat{M}(2\pi j + y) \ne 0$. We show that the latter is equivalent to the nonvanishing of $\widehat{M}$ on Ω^-.

By the definition of Ω, $\widehat{M}$ cannot have a zero in Ω. If there is a $y \in \Omega^-$ with $\widehat{M}(y) = 0$, then there is a sequence $\{y_\nu\}$ in Ω with $y_\nu \to y$, and, for any $j \in \mathbb{Z}^s$,

$$|\widehat{M}(2\pi j + y)| = \lim_{\nu\to\infty} |\widehat{M}(2\pi j + y_\nu)| \le \lim_{\nu\to\infty} |\widehat{M}(y_\nu)| = |\widehat{M}(y)| = 0.$$

On the other hand, suppose there is a $y \in [0\,..\,2\pi)^s$ for which $\widehat{M}(2\pi j + y) = 0$ for all $j \in \mathbb{Z}^s$. Since none of these translates $2\pi j + y$ can be in Ω, each must be in some translate of $\Omega^-\backslash\Omega$. In particular, $y \in (\Omega^-\backslash\Omega) - 2\pi j_0$ for some j_0, and $y + 2\pi j_0$ is a zero of $\widehat{M}$ in $\Omega^-\backslash\Omega$.

□

The boundary $\partial\Omega$ of the set Ω consists of pieces of the real analytic surfaces $f_j - f_0 = 0$ mentioned in the proof of (4)Proposition. The set J_Ω of those $j \in \mathbb{Z}^s\backslash 0$ for which the corresponding surface contributes a piece of positive $(s-1)$–dimensional volume to $\partial\Omega$ enjoys the same symmetries as does Ω: J_Ω is symmetric through the origin and $Q^T J_\Omega = J_\Omega$ if $\Xi \cup -\Xi$ is invariant under the unimodular matrix $Q \in \mathbb{Z}^{s\times s}$. In order to compute Ω, it is important to know that J_Ω is finite. When Ξ is unimodular, the finiteness of J_Ω follows from the fact that $\widehat{M}$ is bounded away from 0 on Ω^-, the boundedness of Ω, and the decay (5).

We now turn to some examples that illustrate the properties given above.

(15)Example: two-direction mesh. Already the simplest situation, viz. $s = 2 = n$ and M is the characteristic function of a parallelogram, provides a surprising richness of fundamental domains. In fact, since Ω_Ξ is recovered from $\Omega_{Q\Xi}$ via (13) when $Q \in \mathbb{Z}^{2\times 2}$ is a unimodular matrix, we may consider them equivalent domains for comparison purposes. This allows Ξ to be reduced to a simpler form:

$$Q\Xi = \left[\begin{smallmatrix}\mu & 0\\ \nu & \varepsilon\end{smallmatrix}\right] \quad \text{with } \mu := |\det \Xi|/\varepsilon,\ \varepsilon := \gcd(\xi_2(1), \xi_2(2)) \text{ and } \nu \in [0\,..\,\mu).$$

The unimodular matrix $Q =: [\zeta_1\ \zeta_2]^T \in \mathbb{Z}^{2\times 2}$ can be chosen as follows: Choose $\zeta_1 = \pm(\xi_2(2), -\xi_2(1))/\varepsilon$ with the appropriate sign so that $\zeta_1\xi_1 = |\det\Xi|/\varepsilon = \mu$. Then $\zeta_1\xi_2 = 0$. Since $\varepsilon = \gcd(\xi_2(1), \xi_2(2))$, there is an integer vector k such that $\xi_2 k = \varepsilon$. Finally, take $\zeta_2 = m\zeta_1 + k$ where the integer m is selected to make $\zeta_2\xi_1 = \nu \in [0\,..\,\mu)$. The matrix Q is unimodular since $\det\Xi = \det Q\Xi$.

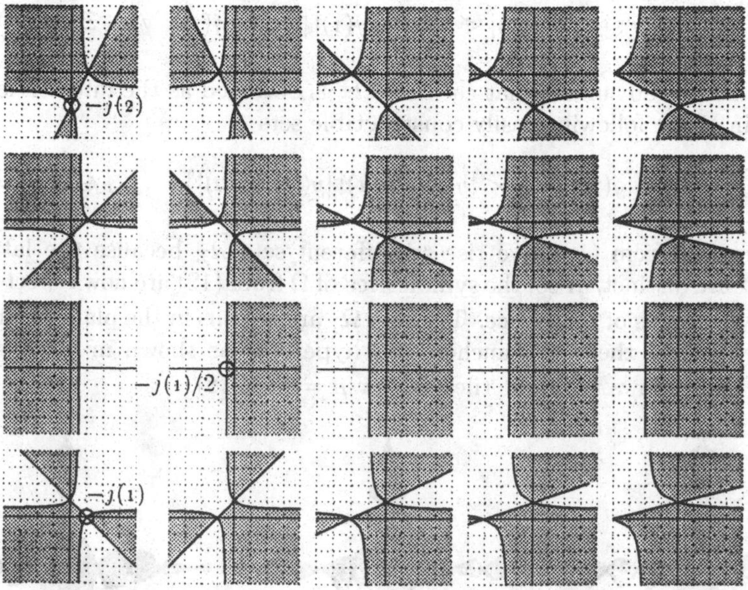

(16)Figure. $\widetilde{\Omega}_j$ for $j \in \{-1, 1, 2, 3, 4\} \times \{-1, 0, 1, 2\}$.

Let H_ξ be the line $\xi y = 0$. From the formula (I.17) for the Fourier transform and (2), we find

(17)
$$\Omega_\Xi\backslash(H_{\xi_1} \cup H_{\xi_2}) = \{y : |y\xi_1|\,|y\xi_2| < |(y+2\pi j)\xi_1|\,|(y+2\pi j)\xi_2|,\ \forall j \in \mathbb{Z}^2\backslash 0\}.$$

In this form it is clear that the open set Ω is not changed if a column $\xi = t\xi^* \in \Xi$ with $\xi^* \in \mathbb{Z}^2$ is replaced by ξ^*. This means that we may further restrict our attention to matrices of the form

$$\left[\begin{smallmatrix}\mu & 0\\ \nu & 1\end{smallmatrix}\right] \quad \text{with} \quad \gcd(\mu,\nu)=1 \quad \text{and} \quad \nu \in [0\,..\,\mu).$$

Finally, since $\left[\begin{smallmatrix}1 & 0\\ 1 & -1\end{smallmatrix}\right]\left[\begin{smallmatrix}\mu & 0\\ \nu & 1\end{smallmatrix}\right] = \left[\begin{smallmatrix}\mu & 0\\ \mu-\nu & -1\end{smallmatrix}\right]$, the set of matrices may be reduced to those of the form

$$(18) \qquad \left[\begin{smallmatrix}\mu & 0\\ \nu & 1\end{smallmatrix}\right] \quad \text{with} \quad \gcd(\mu,\nu)=1 \quad \text{and} \quad \nu \in [0\,..\,\mu/2].$$

For each $\mu < 5$, there is only one matrix in this reduced set, while $\mu = 7$ is the first integer for which there are three.

In order to see that different matrices of the type (18) lead to distinct sets Ω, it is convenient to make the change of variable

$$\Xi^T y/2\pi \mapsto y$$

in (17) to obtain

$$(19) \quad \widetilde{\Omega}_\Xi := \{y : |y(1)|\,|y(2)| < |y(1)+j(1)|\,|y(2)+j(2)|,\ \forall j \in \Xi^T\mathbb{Z}^2\backslash 0\}.$$

This has the advantage that each set $\widetilde{\Omega}_\Xi$ is obtained as the intersection of certain of the following easily constructible sets

$$\widetilde{\Omega}_j := \{y : |y(1)|\,|y(2)| < |y(1)+j(1)|\,|y(2)+j(2)|\}, \qquad j \in \mathbb{Z}^2\backslash 0.$$

Different Ξ from (18) give rise to different sets $\widetilde{\Omega}_\Xi$ because the lattices $\Xi^T\mathbb{Z}^2$ are distinct. Also, the symmetries of Ω from (12) are now reflected in the symmetries of the lattice. The sets $\widetilde{\Omega}_j$ are regions in the plane bounded by a line and the two branches of a hyperbola as shown in (16)Figure. (20)Figure shows the three tiles $\widetilde{\Omega}_\Xi$ for $\mu = 9$ in (18).

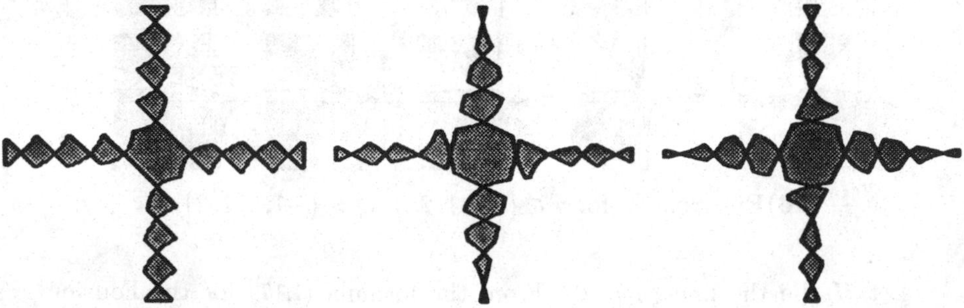

(20)Figure. $\widetilde{\Omega}_\Xi$ for $\Xi = \left[\begin{smallmatrix}9 & 0\\ 1 & 1\end{smallmatrix}\right]$, $\left[\begin{smallmatrix}9 & 0\\ 2 & 1\end{smallmatrix}\right]$, $\left[\begin{smallmatrix}9 & 0\\ 4 & 1\end{smallmatrix}\right]$.

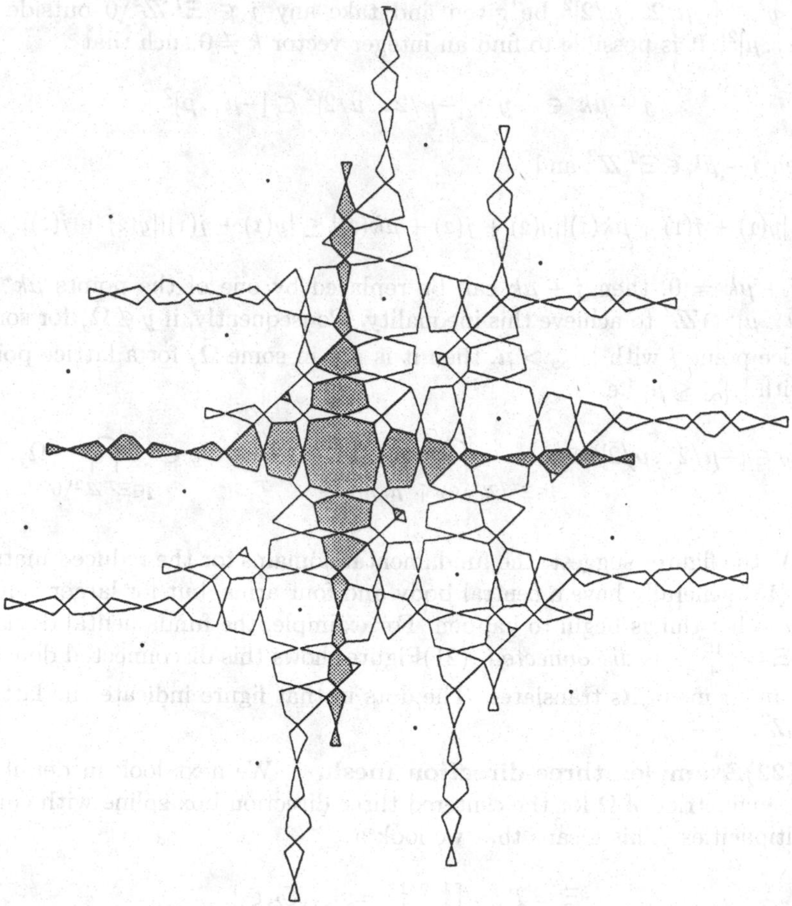

(21)Figure. The disconnected fundamental domain $\widetilde{\Omega}_{\left[\begin{smallmatrix}15 & 0\\ 4 & 1\end{smallmatrix}\right]}$.

The sets $\widetilde{\Omega}_j$ for j on the positive axes are the appropriate half-spaces

$$\{y : y(1) > -j(1)/2\} \quad \text{or} \quad \{y : y(2) > -j(2)/2\}.$$

From $\left[\begin{smallmatrix}\mu & \nu\\ 0 & 1\end{smallmatrix}\right]j = (\mu j(1) + \nu j(2), j(2))$, we see that elements of the lattice are of the form $(\mu j(1), 0)$ on one axis and of the form $(0, -\mu j(1)/\nu)$ on the other. Consequently, $\gcd(\mu, \nu) = 1$ implies that $\pm(\mu, 0)$ and $\pm(0, \mu)$ are the lattice points on the axes closest to 0. Therefore, as the intersection of all $\widetilde{\Omega}_j$ for j in the lattice, $\widetilde{\Omega}_\Xi$ must be confined to the square $[-\mu/2\,..\,\mu/2]^2$.

Finally, we show that $\widetilde{\Omega}_\Xi$ is already the intersection of those Ω_j for which

$$j \in (\Xi^T \mathbb{Z}^2 \backslash 0) \cap [-\mu\,..\,\mu]^2.$$

Let $y \in [-\mu/2 \,..\, \mu/2]^2$ be given and take any $j \in \Xi^T \mathbb{Z}^2 \backslash 0$ outside of $[-\mu \,..\, \mu]^2$. It is possible to find an integer vector $k \neq 0$ such that

$$j + \mu k \in -y + [-\mu/2 \,..\, \mu/2]^2 \subset [-\mu \,..\, \mu]^2.$$

Then $j + \mu k \in \Xi^T \mathbb{Z}^2$ and

$$|y(1) + j(1) + \mu k(1)||y(2) + j(2) + \mu k(2)| \leq |y(1) + j(1)||y(2) + j(2)|.$$

If $j + \mu k = 0$, then $j + \mu k$ can be replaced by one of the points $\mu k^* \in [-\mu \,..\, \mu]^2 \cap \mathbb{Z}^2$ to achieve this inequality. Consequently, if $y \notin \widetilde{\Omega}_j$ for some lattice point j with $|j|_\infty > \mu$, then it is not in some $\widetilde{\Omega}_j$ for a lattice point j with $|j|_\infty \leq \mu$; i.e.,

$$y \in [-\mu/2 \,..\, \mu/2]^2 \cap \bigcap_{j \in \Xi^T(\mathbb{Z}^2 \backslash 0) \cap [-\mu..\mu]^2} \widetilde{\Omega}_j \implies y \in \bigcap_{j \in \Xi^T \mathbb{Z}^2 \backslash 0} \widetilde{\Omega}_j.$$

As the figures suggest, the fundamental domains for the reduced matrix set (18) generally have a central body and four arms, but for larger values of μ, other things begin to happen. For example, the fundamental domain for $\Xi = \left[\begin{smallmatrix} 15 & 0 \\ 4 & 1 \end{smallmatrix}\right]$ is *disconnected.* (21)Figure shows this disconnected domain $\widetilde{\Omega}_\Xi$ and some of its translates. The dots in that figure indicate the lattice $\Xi^T \mathbb{Z}^2$. ◻

(22)Example: three-direction mesh. We next look in detail at the symmetries of Ω for the centered three-direction box spline with equal multiplicities. This means that we look at

$$\Xi = T := \left[\begin{smallmatrix} 1 & 0 & 1 \\ 0 & 1 & 1 \end{smallmatrix}\right] =: [\zeta_1, \zeta_2, \zeta_3].$$

The unimodular operators that leave the matrix T invariant up to the signs of its columns are readily described in terms of the symmetric group $\mathcal{S}_3$ of order 3. Each such operator is determined by its action $\zeta_\nu \mapsto \pm\zeta_{\sigma(\nu)}$, $\nu = 1, 2, 3$, where $\sigma \in \mathcal{S}_3$ and some sign pattern appears. For each $\sigma \in \mathcal{S}_3$, there are exactly two such unimodular matrices associated with σ. If we denote one of the two by Q_σ, then the other is $-Q_\sigma =: Q_{-\sigma}$. For example, we could choose

(23) $$Q_{(1,2)} = \left[\begin{smallmatrix} 0 & -1 \\ -1 & 0 \end{smallmatrix}\right], \quad Q_{(1,3)} = \left[\begin{smallmatrix} 1 & 0 \\ 1 & -1 \end{smallmatrix}\right],$$

and these generate the set

$$\mathbf{Q} := \{Q_{\pm\sigma} : \sigma \in \mathcal{S}_3\}$$

of all relevant unimodular matrices by the rules $Q_{\sigma\tau} := Q_\sigma Q_\tau$, $Q_{-\sigma} := -Q_\sigma$. In these terms, the symmetries of the centered three-direction box spline with equal multiplicities are given by $Q \in \mathbf{Q}$ in the sense that (see (I.23) and (I.24)) $M^c_{r,r,r} = M^c_{r,r,r} \circ Q$ for all such Q. Thus, the full set of symmetries for $M^c_{r,r,r}$ is generated by the symmetry through the origin and the symmetries given by the matrices (23),

$$M^c_{r,r,r}(x(1), x(2)) = M^c_{r,r,r}(-x(2), -x(1)) = M^c_{r,r,r}(x(1), x(1) - x(2)).$$

According to (12), the symmetries of Ω are given by the set

$$\mathbf{Q}^T := \{Q^T_\sigma : Q_\sigma \in \mathbf{Q}\}.$$

The full set of symmetries on Ω obtained in this way are

$$\begin{aligned}\pm(y(1), y(2)) \in \Omega \Longrightarrow \{ &\pm(y(2), y(1)), \pm(y(1) + y(2), -y(2)),\\ &\pm(-y(1), y(1) + y(2)), \pm(y(1) + y(2), -y(1)),\\ &\pm(-y(2), y(1) + y(2))\} \in \Omega.\end{aligned}$$

With the choice (23) of generators, the action of $\mathbf{Q}^T$ on $\mathbb{R}^2$ can be described by

(24) $$Q^T_\sigma C = C_\sigma, \qquad \sigma \in \mathcal{S}_3$$

where

$$C := \{y \in \mathbb{R}^2 : \min\{y(1), y(2)\} \geq 0\}$$

is the positive quadrant. The six cones C_σ are shown in (25)Figure.

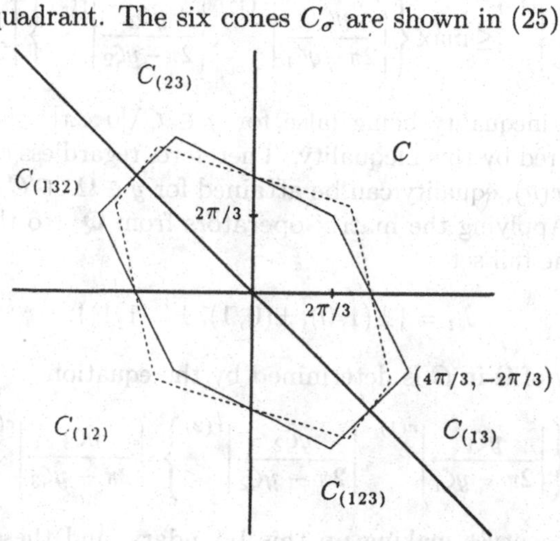

(25)Figure. Cones C_σ of symmetry, and Ω_{T_r} for $r = (t, t, t)$ and $r = (7, 5, 2)$.

The symmetry group $\mathbf{Q}^T$ is useful in the description of Ω for any of the three-direction matrices T_r whatever the nonzero multiplicities. For multiplicities $r = (r(1), r(2), r(3))$, let σr be this set of integers permuted according to σ. Since $Q_\sigma T_r = T_{\sigma r}$, (13) reads

$$\Omega_{T_r} = Q_\sigma^T \Omega_{T_{\sigma r}}. \tag{26}$$

Combining this relation with (24), we find that Ω is completely determined for any of the matrices T_r once it is described on the cone C for every T_r.

The final description of Ω rests with finding the set J_Ω of those j that determine the boundary of $\Omega \cap C$. For $y \in C \backslash 0$, the definition of $y \in \Omega$ requires

$$\left| \frac{\widehat{M}^c_{r(1),r(2),r(3)}(2\pi j + y)}{\widehat{M}^c_{r(1),r(2),r(3)}(y)} \right| = \prod_\nu \left| \frac{y\zeta_\nu}{(2\pi j + y)\zeta_\nu} \right|^{r(\nu)} < 1, \qquad j \in \mathbb{Z}^s \backslash 0, \tag{27}$$

while for finding J_Ω, we are particularly interested in when equality holds here. For this, we observe that for real $w > 0$ and real p

$$\left| \frac{w}{|p| + w} \right| < \left| \frac{w}{|p| - w} \right|$$

and that the second quantity is greater than 1 if $w > |p|/2$. Therefore, each factor in (27) will be largest if $j\zeta_\nu$ is negative. If y is further restricted to the square $[0\,..\,\pi]^2$, then each factor is largest when the $|j\zeta_\nu|$ are the smallest possible. Thus, for $y \in \Omega \cap C \cap [0\,..\,\pi]^2$,

$$\prod_\nu \left| \frac{y\zeta_\nu}{(2\pi j + y)\zeta_\nu} \right|^{r(\nu)} \le \max\left\{ \left| \frac{y\zeta_1}{2\pi - y\zeta_1} \right|^{r(1)}, \left| \frac{y\zeta_2}{2\pi - y\zeta_2} \right|^{r(2)} \right\} \left| \frac{y\zeta_3}{2\pi - y\zeta_3} \right|^{r(3)} < 1,$$

while the last inequality being false for $y \in C \backslash [0\,..\,\pi]^2$ shows that all of $\Omega \cap C$ is captured by this inequality. Therefore, regardless of the (nonzero) multiplicities $r(\nu)$, equality can be attained for $y \in \Omega^- \cap C$ only if $j = -\zeta_1$ or $j = -\zeta_2$. Applying the matrix operators from $\mathbf{Q}^T$ to these two values of j, we get the full set

$$J_\Omega = \{\pm(1,0), \pm(0,1), \pm(-1,1)\}.$$

The boundary of Ω in C is determined by the equation

$$\max\left\{ \left| \frac{y\zeta_1}{2\pi - y\zeta_1} \right|^{r(1)}, \left| \frac{y\zeta_2}{2\pi - y\zeta_2} \right|^{r(2)} \right\} \left| \frac{y\zeta_3}{2\pi - y\zeta_3} \right|^{r(3)} = 1.$$

There are two curves making up this boundary, and these reduce in the case of equal multiplicities to the two lines

$$y(1) + 2y(2) = 2\pi, \qquad 2y(1) + y(2) = 2\pi.$$

These two lines intersect in the point $(2\pi/3, 2\pi/3)$. Applying the operators in $\mathbf{Q}^T$ to this point, we find that the fundamental domain for the three-direction box spline with equal multiplicities is the convex hull of the points

$$(28) \qquad \pm(2\pi/3, 2\pi/3), \quad \pm(4\pi/3, -2\pi/3), \quad \pm(2\pi/3, -4\pi/3).$$

The outline of this domain is shown in (25)Figure. In general, the fundamental domain for a three-direction box spline with nonzero multiplicities has a hexagonal shape, although the sides are not straight lines except in the case of equal multiplicities, and the domain is not convex in general. The outline of the fundamental domain corresponding to the matrix $T_{7,5,2}$ is dashed in (25)Figure. ◻

Convergence of cardinal splines. We now return to the topic that motivated the discovery of Ω: The convergence, as the degree goes to infinity, of a sequence of cardinal splines in several variables. It can be characterized in terms of Fourier transforms, so it is natural to use $\mathbf{L}_2$-norms and to restrict the coefficients of the box spline shifts to $\ell_2(\mathbb{Z}^s)$. Define

$$S_2(M_{*_r}) := \{M_{*_r} * a : a \in \ell_2(\mathbb{Z}^s)\}.$$

Since M_{*_r} is bounded and has compact support, $S_2(M_{*_r}) \subset \mathbf{L}_2(\mathbb{R}^s)$, and consequently, $\widehat{S_2(M_{*_r})} \subset \mathbf{L}_2(\mathbb{R}^s)$. With

$$\operatorname{dist}(f, S_2(M_{*_r})) := \inf\{\|f - g\|_2 : g \in S_2(M_{*_r})\}$$

the distance of a function $f \in \mathbf{L}_2(\mathbb{R}^s)$ to $S_2(M_{*_r})$, we wish to characterize the class

$$S_2(M_{*_\infty}) := \{f \in \mathbf{L}_2(\mathbb{R}^s) : \lim_{r\to\infty} \operatorname{dist}(f, S_2(M_{*_r})) = 0\}.$$

(29)Theorem. *For $M = M_\Xi$ with $\operatorname{ran}\Xi = \mathbb{R}^s$, a function $f \in \mathbf{L}_2(\mathbb{R}^s)$ is in $S_2(M_{*_\infty})$ if and only if the support of $\widehat{f}$ is contained in Ω^-.*

Proof. We need several auxiliary estimates which follow from the decay of $\widehat{M}$ and the definition of Ω. Let

$$N := \{y \in \mathbb{R}^s : \widehat{M}(y) \neq 0\}.$$

Then $\Omega \subset N$, and $\mathbb{R}^s \backslash N$ is a set of measure zero. Therefore, the functions

$$h_j := \widehat{M}(\cdot + 2\pi j)/\widehat{M}, \qquad j \in \mathbb{Z}^s,$$

are defined almost everywhere in $\mathbb{R}^s$ and everywhere on Ω. Using Poisson's summation formula and the relation $\widehat{M_{*_r}} = (\widehat{M})^r$, the symbol $\widetilde{M}_{*_r}$ can be written on N as

$$\begin{aligned}\widetilde{M}_{*_r}(y) &= \sum_{j\in\mathbb{Z}^s} M_{*_r}(j)\exp(-ijy) = \sum_{j\in\mathbb{Z}^s} \widehat{M}_{*_r}(y+2\pi j)\\ &= \widehat{M}_{*_r}(y)\sum_{j\in\mathbb{Z}^s}(h_j(y))^r.\end{aligned}$$

Let $y\in\Omega$. For any $j\neq 0$, there exists $\varepsilon(j,y)>0$ such that

$$(30)\qquad |h_j(y)| \le 1-\varepsilon(j,y) < 1, \qquad y\in\Omega,$$

while by (8) there exists a positive const (depending on y) such that for all but finitely many j

$$(31)\qquad |h_j(y)| \le 1/(1+\text{const}|j|), \qquad y\in\Omega.$$

Consequently,

$$(32)\qquad \widetilde{M}_{*_r}(y)/\widehat{M}_{*_r}(y) = \sum_{j\in\mathbb{Z}^s}(h_j(y))^r \to 1, \qquad r\to\infty, \quad y\in\Omega.$$

Moreover, the convergence in (32) and the bounds in (30) and (31) are uniform on compact subsets Ω_1 of Ω. In particular, $\widetilde{M}_{*_r}$ does not vanish on a compact set $\Omega_1\subset\Omega$ if r is sufficiently large, i.e., $r\ge r(\Omega_1)$.

Suppose now that $f\in \mathbf{L}_2(\mathbb{R}^s)$ and that $\widehat{f}$ has support in Ω^-. An approximating sequence to f,

$$g_r := M_{*_r} * a_r \in S_2(M_{*_r}),$$

is constructed as follows. Let Ω_1 be a compact subset of Ω and let χ denote its characteristic function. Since Ω is a fundamental domain, $\widehat{f}\chi/\widetilde{M}_{*_r}$ can be expanded in a Fourier series

$$(\widehat{f}\chi/\widetilde{M}_{*_r})(y) =: \sum_{j\in\mathbb{Z}^s}\exp(-ijy)a_r(j), \qquad y\in\Omega.$$

Let $(\widehat{f}\chi/\widetilde{M}_{*_r})^\circ$ be its periodic extension to $\cup_{j\in\mathbb{Z}^s}(\Omega+2\pi j)$ as defined by the Fourier series. We claim that $\widehat{g}_r = (\widehat{f}\chi/\widetilde{M}_{*_r})^\circ\widehat{M}_{*_r}$ converges to the Fourier transform of f in $\mathbf{L}_2(\mathbb{R}^s)$ as $r\to\infty$ and $\Omega_1\to\Omega^-$. Indeed, since $\widehat{f}$ vanishes a.e. outside Ω^- and $\Omega^-\backslash\Omega$ has measure zero by (4)Proposition, we have

$$(33)\qquad \|\widehat{f}-\widehat{g}_r\|_2^2 = \|\widehat{f}-\widehat{g}_r\|_{2,\Omega}^2 + \sum_{j\in\mathbb{Z}^s\backslash 0}\|\widehat{g}_r(\cdot+2\pi j)\|_{2,\Omega}^2.$$

The first term can be estimated by

$$\|\widehat{f} - \widehat{g}_r\|_{2,\Omega} \le \|\widehat{f} - \chi\widehat{f}\|_{2,\Omega} + \|\chi\widehat{f} - \chi\widehat{f}\widehat{M}_{*_r}/\widetilde{M}_{*_r}\|_{2,\Omega},$$

where the first term on the right-hand side is small if Ω_1 is chosen close to Ω, while, for fixed Ω_1, the second term is small by (32) if r is sufficiently large. The jth term in the sum in (33) is the square of

$$\begin{aligned}\|\widehat{M}_{*_r}(\cdot + 2\pi j)(\widehat{f}\chi/\widetilde{M}_{*_r})\|_{2,\Omega} &= \|h_j^r \widehat{M}_{*_r}(\widehat{f}\chi/\widetilde{M}_{*_r})\|_{2,\Omega} \\ &\le \|h_j\|_{\infty,\Omega_1}^r \|\widehat{M}_{*_r}/\widetilde{M}_{*_r}\|_{\infty,\Omega_1} \|\widehat{f}\|_{2,\Omega_1}.\end{aligned}$$

By (30), (31) and (32), this estimate implies that the sum in (33) is small for large r. Hence,

$$\|g_r - f\|_2 = \frac{1}{(2\pi)^s}\|\widehat{g}_r - \widehat{f}\|_2 \to 0 \qquad \text{as} \quad r \to \infty,$$

by Parseval's identity.

On the other hand, suppose that, for some $a_r \in \ell_2$, $g_r = M_{*_r} * a_r \in S_2(M_{*_r})$ converges to the function f in $\mathbf{L}_2(\mathbb{R}^s)$. Since $\widehat{g}_r = \widehat{M}_{*_r} H_r$, with $H_r := \sum_{j\in\mathbb{Z}^s} \exp(-ij\cdot)a_r(j)$ 2π-periodic, we have

$$\widehat{g}_r(y + 2\pi j) = (h_j(y))^r \widehat{g}_r(y), \qquad y \in N.$$

Consequently, by (30) and (31), for any compact subset Ω_1 of Ω and all $j \in \mathbb{Z}^s\backslash 0$,

$$\|\widehat{g}_r\|_{2,\Omega_1+2\pi j} \le \|h_j\|_{\infty,\Omega_1}^r \|\widehat{g}_r\|_{2,\Omega} \xrightarrow[r\to\infty]{} 0.$$

Since $\widehat{g}_r$ converges to $\widehat{f}$ in $\mathbf{L}_2(\mathbb{R}^s)$ and $\mathbb{R}^s\backslash \cup_j (\Omega + 2\pi j)$ has measure zero, it follows that $\widehat{f}$ vanishes almost everywhere outside Ω^-. □

(34)Corollary. *If cardinal interpolation with M_{*_r} is correct for every r, then, for any $f \in \mathbf{L}_2(\mathbb{R}^s)$ with $\operatorname{supp}\widehat{f} \subset \Omega$, the cardinal interpolants, $\mathcal{L}_r f$, converge to f in $\mathbf{L}_2(\mathbb{R}^s) \cap \mathbf{L}_\infty(\mathbb{R}^s)$. In fact, if $\operatorname{supp}\widehat{f} \subseteq \Omega_1$ for some compact subset Ω_1 of Ω, then*

$$\max\{\|f - \mathcal{L}_r f\|_2, \|f - \mathcal{L}_r f\|_\infty\} \le \operatorname{const}(r, \Omega_1)\|f\|_2$$

with $\operatorname{const}(r, \Omega_1) \to 0$ *as* $r \to \infty$.

Proof. If cardinal interpolation with M_{*_r} is correct then, by (IV.12) Theorem, the symbol $\widetilde{M}_{*_r}$ does not vanish and the coefficients of the fundamental function L are given by (IV.9). Therefore,

$$\widehat{L}_r = \sum_{j\in\mathbb{Z}^s} \exp(-ij\cdot)(1/\widetilde{M}_{*_r})^\vee(j)\widehat{M}_{*_r} = \widehat{M}_{*_r}/\widetilde{M}_{*_r},$$

and

$$(35) \quad \Big(\sum_{|j|\le j_0} L_r(\cdot - j) f(j)\Big)\widehat{\ } = \sum_{|j|\le j_0} f(j)\exp(-ij\cdot)\widehat{M}_{*_r}/\widetilde{M}_{*_r}, \qquad \forall j_0.$$

Since Ω is a fundamental domain and $\widehat{f} \in \mathbf{L}_2(\mathbb{R}^s)$ with support in Ω, we have

$$f(j) = \frac{1}{(2\pi)^s}\int_\Omega \widehat{f}(y)\exp(ijy)\,dy,$$

are Fourier coefficients of $\widehat{f}^\circ$, the periodic extension of $\widehat{f}$, i.e.,

$$\widehat{f}^\circ(y + 2\pi j) = \widehat{f}(y), \quad y \in \Omega, j \in \mathbb{Z}^s.$$

Then, letting $j_0 \to \infty$ in (35), shows that

$$\widehat{\mathcal{L}_r f} = \widehat{f}^\circ \widehat{M}_{*_r}/\widetilde{M}_{*_r}.$$

With $\widehat{f}^\circ \widehat{M}_{*_r}/\widetilde{M}_{*_r}$ replacing $\widehat{g}_r$, the proof for convergence in the $\mathbf{L}_2(\mathbb{R}^s)$ norm follows as in (29)Theorem. The same argument can be modified for the $\mathbf{L}_\infty(\mathbb{R}^s)$ norm:

(36)

$$\begin{aligned}
\|\mathcal{L}_r f - f\|_\infty &\le \frac{1}{(2\pi)^s}\|\widehat{f} - \widehat{f}^\circ \widehat{M}_{*_r}/\widetilde{M}_{*_r}\|_1 \\
&= \frac{1}{(2\pi)^s}\|\widehat{f} - \widehat{f}\widehat{M}_{*_r}/\widetilde{M}_{*_r}\|_{1,\Omega} + \sum_{j\in\mathbb{Z}^s\backslash 0}\frac{1}{(2\pi)^s}\|\widehat{M}_{*_r}(\cdot + 2\pi j)\widehat{f}/\widetilde{M}_{*_r}\|_{1,\Omega} \\
&\le \frac{1}{(2\pi)^{s/2}}\Big(\|\widehat{f} - \widehat{f}\widehat{M}_{*_r}/\widetilde{M}_{*_r}\|_{2,\Omega_1} + \sum_{j\in\mathbb{Z}^s\backslash 0}\|\widehat{M}_{*_r}(\cdot + 2\pi j)\widehat{f}/\widetilde{M}_{*_r}\|_{2,\Omega_1}\Big) \\
&\le \frac{1}{(2\pi)^{s/2}}\Big(\|1 - \widehat{M}_{*_r}/\widetilde{M}_{*_r}\|_{\infty,\Omega_1} \\
&\qquad + \|\widehat{M}_{*_r}/\widetilde{M}_{*_r}\|_{\infty,\Omega_1}\sum_{j\in\mathbb{Z}^s\backslash 0}(\|h_j\|_{\infty,\Omega_1})^r\Big)\|\widehat{f}\|_{2,\Omega_1}.
\end{aligned}$$

The last terms go to zero as before since Ω_1 is compact in Ω. □

Relation to cardinal series. The sequence of univariate cardinal spline interpolants to a continuous $\mathbf{L}_2(\mathbb{R})$ function f has the Whittaker cardinal series

$$Wf := \sum_{j\in\mathbb{Z}} \operatorname{sinc}(\pi(\cdot - j))f(j)$$

as its limit as the degree tends to infinity. With the notion of degree tending to infinity replaced by Ξ_r, $r \to \infty$, there is still the difficulty that the tensor product definition of cardinal series is associated with the set $(-\pi \,..\, \pi)^s$ and therefore does not accommodate the sets Ω associated with box splines. Several multidimensional analogues of cardinal series have been defined. A reasonably general definition of cardinal series that complements cardinal box splines is the following one: Let $K \in \mathbf{L}_2(\mathbb{R}^s)$ be any function with

(37) $$\sum_{j\in\mathbb{Z}^s} |\widehat{K}(\cdot + 2\pi j)|^2 = \text{const} \quad a.e.$$

Then the functions $\{K(\cdot - j)\}$, $j \in \mathbb{Z}^s$, form an orthogonal system in $\mathbf{L}_2(\mathbb{R}^s)$:

(38)
$$\begin{aligned}\int_{\mathbb{R}^s} K(x-k)\overline{K(x)}\,dx &= \frac{1}{(2\pi)^s}\int_{\mathbb{R}^s} \exp(-iky)|\widehat{K}(y)|^2\,dy \\ &= \frac{1}{(2\pi)^s}\int_{[-\pi..\pi]^s} \sum_{j\in\mathbb{Z}^s} |\widehat{K}(y+2\pi j)|^2 \exp(-iky)\,dy \\ &= \text{const}\,\delta(k).\end{aligned}$$

For an $\mathbf{L}_2$-function f, **the K-cardinal series** for f is defined as

(39) $$W_K f := K{*}a_f,$$

where

(40) $$a_f(j) := \int_{\mathbb{R}^s} f(x)\overline{K(x-j)}\,dx.$$

In the simplest possible case, $\widehat{K} = \chi_\Omega$ a.e., with Ω a fundamental domain for Fourier analysis in $\mathbb{R}^s$; i.e., Ω satisfies (3). In this case, for any function $f \in \mathbf{L}_2(\mathbb{R}^s)$ with $\operatorname{supp}\widehat{f} \subset \Omega$,

$$\begin{aligned} a_f(j) &= \int_{\mathbb{R}^s} f(x)\overline{K(x-j)}\,dx = \frac{1}{(2\pi)^s}\int_{\mathbb{R}^s} \widehat{f}(y)\widehat{K}(y)\exp(ijy)dy \\ &= \frac{1}{(2\pi)^s}\int_\Omega \widehat{f}(y)\exp(ijy)\,dy = f(j).\end{aligned}$$

In particular, $K_{|\mathbb{Z}^s} = \delta$; therefore, the K-cardinal series interpolates such functions f on $\mathbb{Z}^s$. It is known that $W_K f = f$, with the series converging uniformly and in $\mathbf{L}_2(\mathbb{R}^s)$. Therefore, (34)Corollary points out that if cardinal interpolation is correct for M_{*_r}, then the cardinal spline interpolant to f converges in $\mathbf{L}_2(\mathbb{R}^s) \cap \mathbf{L}_\infty(\mathbb{R}^s)$ to $W_K f$ when $\widehat{K} = \chi_\Omega$. In fact, the cardinal spline interpolation operators $\mathcal{L}_r$ converge to the K-cardinal series operator on each of the spaces

$$\{f \in \mathbf{L}_2(\mathbb{R}^s) : \operatorname{supp}\widehat{f} \subseteq \Omega_1\}$$

equipped with the $\mathbf{L}_2(\mathrm{IR}^s)$ norm, where Ω_1 is a compact subset of Ω.

Example: three-direction mesh. The fundamental domain Ω for the three-direction box spline with equal multiplicities shown in (25)Figure can be written in several ways as the union of three parallelograms using the points (28). This makes it relatively easy to find the function K for which $\widehat{K} = \chi_\Omega$: For $\zeta \in \mathrm{IR}^2$, define $\zeta^* := (\zeta(2), -\zeta(1))$. If the two vectors ζ, η determine a parallelogram $R(\zeta,\eta)$ by the lines

$$\zeta^* y = 0, \quad \zeta^* y = \zeta^* \eta \quad \text{and} \quad \eta^* y = 0, \quad \eta^* y = \eta^* \zeta,$$

then by the change of variable $w = (\zeta^* y, \eta^* y)$, we find

$$\begin{aligned}\frac{1}{(2\pi)^2}&\int_{R(\zeta,\eta)} \exp(iyx)\,dy \\ &= \frac{1}{(2\pi)^2}\frac{1}{\eta^*\zeta}\int_{[0..\eta^*\zeta]^2} \exp\left(\frac{i\eta x w(1)}{-\eta^*\zeta} + \frac{i\zeta x w(2)}{\eta^*\zeta}\right) dw(1)\,dw(2) \\ &= \frac{\eta^*\zeta}{(2\pi)^2}\frac{(\exp(i\eta x)-1)(\exp(i\zeta x)-1)}{\eta x\,\zeta x}.\end{aligned}$$

Applying this formula to each of the parallelograms generated by $\zeta_1 = \frac{2\pi}{3}(1,1)$, $\eta_1 = \frac{2\pi}{3}(1,-2)$, $\zeta_2 = \frac{2\pi}{3}(-2,1)$, $\eta_2 = \frac{2\pi}{3}(1,1)$, and $\zeta_3 = \frac{2\pi}{3}(1,-2)$, $\eta_3 = \frac{2\pi}{3}(-2,1)$, and simplifying the result using the fact that $\zeta_1+\zeta_2+\zeta_3 = 0 = \eta_1+\eta_2+\eta_3$, we obtain

$$\begin{aligned}K(x) &= \frac{1}{(2\pi)^2}\int_\Omega \exp(iyx)\,dy \\ &= \frac{-6}{(2\pi)^2}\left[\frac{\cos\frac{2\pi}{3}(x(1)-2x(2))}{(x(2)-2x(1))(x(1)+x(2))} + \frac{\cos\frac{2\pi}{3}(x(1)+x(2))}{(x(1)-2x(2))(x(2)-2x(1))}\right. \\ &\qquad\left. + \frac{\cos\frac{2\pi}{3}(x(2)-2x(1))}{(x(1)-2x(2))(x(1)+x(2))}\right].\end{aligned}$$

□

A more direct connection between cardinal series and cardinal splines occurs when the space $S_2(M)$ is generated by a cardinal spline K in the sense that it consists of all K-cardinal series with coefficients from $\ell_2(\mathbb{Z}^s)$, i.e.,

$$S_2(M) = S_2(K)$$

for some cardinal spline K for which the set $\{K(\cdot - j)\}_{j\in\mathbb{Z}^s}$ of shifts is orthogonal. The key to the construction of such K is the formula (37). For any translate M of the box spline M_Ξ, the function

$$(41) \qquad [\widehat{M},\widehat{M}] : \mathrm{IR}^s \to \mathrm{IR} : y \mapsto \sum_{j\in\mathbb{Z}^s} |\widehat{M}(y+2\pi j)|^2$$

is positive except possibly for a set of measure zero, and since $\widehat{M} = \exp(\cdot i\alpha)\widehat{M^c}$ for some α, $[\widehat{M},\widehat{M}]$ is also the symbol for M^c_{*2}, by the Poisson summation formula; hence it is a trigonometric polynomial. Therefore, the function

$$\widehat{K} := \widehat{M}/[\widehat{M},\widehat{M}]^{1/2} \tag{42}$$

is in $\mathbf{L}_2(\mathrm{I\!R}^s)$ and satisfies (37). The function $K \in \mathbf{L}_2(\mathrm{I\!R}^s)$ defined by the Fourier transform (42) is in the $\mathbf{L}_2$-closure

$$S_2(M)^-$$

of $S_2(M)$. Indeed, by Lebesgue's dominated convergence theorem, $\widehat{K}$ is the $\mathbf{L}_2$-limit of the functions $\widehat{f_r} := \widehat{M}\chi^\circ_{\Omega_r}/[\widehat{M},\widehat{M}]^{1/2}$ where $\chi^\circ_{\Omega_r}$ are periodic extensions (with respect to the fundamental domain Ω_M) of characteristic functions of compact $\Omega_r \subset \Omega_M$, with $\chi_{\Omega_r} \to \chi_{\Omega_M}$ as $r \to \infty$. Each f_r is in $S_2(M)$ since $[\widehat{M},\widehat{M}] > 0$ on Ω_M (cf. the proof of (29)Theorem).

If the shifts of M are linearly independent, or equivalently, Ξ is unimodular, then by (14)Proposition, $[\widehat{M},\widehat{M}] > 0$ on Ω_M^-, and K has the representation

$$K = M{*}a_K \tag{43}$$

with the coefficient sequence a_K given by

$$a_K(j) = a_K(-j) := \frac{1}{(2\pi)^s}\int_{[-\pi..\pi]^s} \frac{\exp(-ijy)}{([\widehat{M},\widehat{M}](y))^{1/2}}\,dy, \qquad j \in \mathbb{Z}^s.$$

The mapping

$$K{*}\colon \mathrm{I\!R}^{\mathbb{Z}^s} \to \mathrm{I\!R}^{\mathrm{I\!R}^s} : a \mapsto K{*}a := \sum_{j\in\mathbb{Z}^s} K(\cdot - j)a(j)$$

carries the sequence $a = \{a_f(j)\}_{j\in\mathbb{Z}^s}$ given by (40) to the K-cardinal series $W_K f$; this is the so-called K-cardinal spline for f.

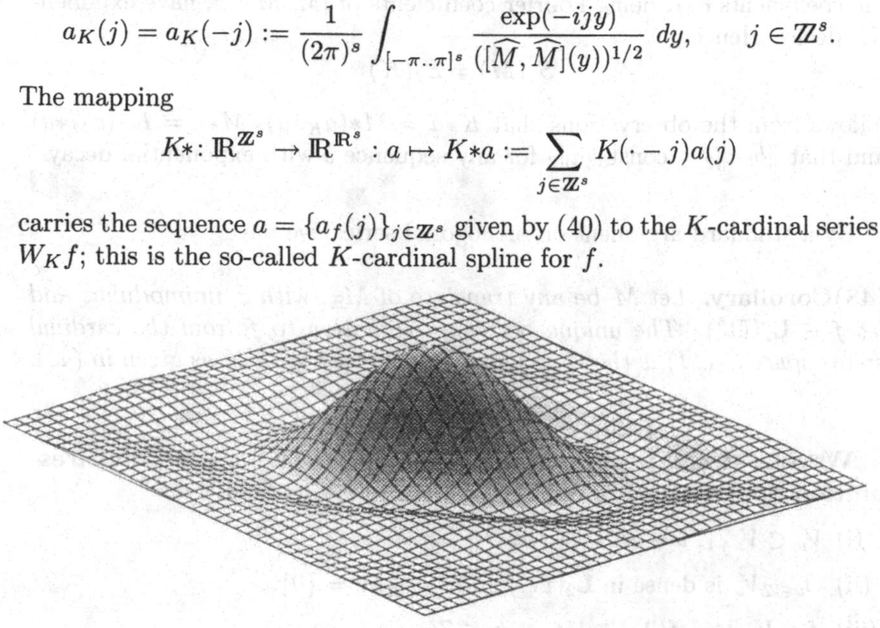

(44)Figure. The cardinal spline K for $M^c_{2,2,2}$ on $[-2.5\,..\,2.5]^2$.

(45)Theorem. *If M is any translate of M_Ξ, then the shifts $\{K(\cdot-j)\}_{j\in\mathbb{Z}^s}$ of the function K defined by (42) form an orthonormal basis for $S_2(M)^-$. Moreover, if Ξ is unimodular, then $S_2(M)^- = S_2(M)$, K has the representation (43), and*

(46)
$$|a_K(j)| \le \text{const}\exp(-|j|/\text{const}) \quad \textit{and} \quad |K(x)| \le \text{const}\exp(-|x|/\text{const}),$$

for some positive constants; in particular, for any $p \in [1\,..\,\infty]$ and $a \in \ell_p(\mathbb{Z}^s)$,

$$\|K*a\|_p \le \text{const}\|a\|_p. \tag{47}$$

Proof. The fact that $\{K(\cdot - j)\}_{j\in\mathbb{Z}^s}$ is an orthonormal system for $S_2(M)^-$ is a consequence of (37) and (38) as discussed earlier.

When Ξ is unimodular, the proofs of (46) and (47) are the same as the proofs of (IV.13) and (IV.14) in (IV.12)Theorem with $1/\widetilde{M}$ replaced by $1/[\widehat{M},\widehat{M}]^{1/2}$, since $1/[\widehat{M},\widehat{M}]^{1/2}$, being the reciprocal of the square root of a nonvanishing trigonometric polynomial, is analytic in a neighborhood of $[-\pi\,..\,\pi]^s$ in $\mathbf{C}^s$.

Similarly, since $\widehat{M} = \widehat{K}[\widehat{M},\widehat{M}]^{1/2}$, we conclude that $M = K*a_M$, where the coefficients a_M, being Fourier coefficients of $[\widehat{M},\widehat{M}]^{1/2}$, have exponential decay. Hence,

$$S_2(M) = S_2(K)$$

follows from the observations that $K*a = M*(a_K*a)$, $M*a = K*(a_M*a)$ and that $\|b*a\|_2 \le \text{const}_b\|a\|_2$ for any sequence b with exponential decay. □

By a standard argument on orthogonal series, we have

(48)Corollary. *Let M be any translate of M_Ξ, with Ξ unimodular, and let $f \in \mathbf{L}_2(\mathbb{R}^s)$. The unique best approximation to f from the cardinal spline space $S_2(M)$ is the K-cardinal spline $W_K f$ with K as given in (43).*

Wavelet decompositions. A scale of spaces $\{V_\nu\}_{\nu\in\mathbb{Z}}$ is a **multiresolution approximation** of $\mathbf{L}_2(\mathbb{R}^s)$ if the following hold:

(i) $V_\nu \subset V_{\nu+1}$, $\nu \in \mathbb{Z}$.

(ii) $\cup_{\nu\in\mathbb{Z}} V_\nu$ is dense in $\mathbf{L}_2(\mathbb{R}^s)$ and $\cap_{\nu\in\mathbb{Z}} V_\nu = \{0\}$.

(iii) $f \in V_\nu \iff f(2\,\cdot\,) \in V_{\nu+1}$, $\forall \nu \in \mathbb{Z}$.

(iv) There is an isomorphism from V_0 onto $\ell_2(\mathbb{Z}^s)$ which commutes with all shifts, i.e., all translations τ_j, $j \in \mathbb{Z}^s$.

The spaces V_ν of interest here are

$$(49)\qquad V_\nu := \{\sum_{j\in\mathbb{Z}^s} M(2^\nu\cdot -j)a(j) : a\in \ell_2(\mathbb{Z}^s)\}^-,$$

where $M = M_\Xi$.

(50)Theorem. *When $M = M_\Xi$, the spaces V_ν, $\nu\in\mathbb{Z}$, defined in (49) form a multiresolution approximation of $\mathbf{L}_2(\mathbb{R}^s)$.*

Proof. For (i), it suffices to show that the generator $M(\cdot/2)/2^s$ of V_{-1} belongs to V_0. This follows since (see (I.17))

$$\frac{\widehat{M}(2y)}{\widehat{M}(y)} = \prod_{\xi\in\Xi}\frac{1+\exp(-i\xi y)}{2}$$

is 2π-periodic, and consequently, $M(\cdot/2)/2^s = M{*}m$ for some finitely supported mask m. This observation is the basis for subdivision algorithms and will be discussed in more detail in Chapter VII.

(ii) The first assertion follows from (III.4)Proposition. For the second assertion, it suffices to prove that, with $P_\nu : \mathbf{L}_2(\mathbb{R}^s)\to V_\nu$ the orthogonal projector onto V_ν, $P_\nu f \to 0$ as $\nu\to -\infty$ for any compactly supported $f\in\mathbf{L}_2(\mathbb{R}^s)$ (since these are dense in $\mathbf{L}_2(\mathbb{R}^s)$). By (45)Theorem, the shifts $\{2^{\nu s/2}K(2^\nu\cdot -j)\}_{j\in\mathbb{Z}^s}$ form a complete orthonormal system for V_ν; hence, for any $f\in\mathbf{L}_2(\mathbb{R}^s)$,

$$P_\nu f = \sum_{j\in\mathbb{Z}^s} K(2^\nu\cdot -j)a_{f,\nu}(j).$$

If also f is supported in the cube $[-R\,..\,R]^s$, we have, for all ν near $-\infty$,

$$\begin{aligned}\|P_\nu f\|_2^2 &= \sum_{j\in\mathbb{Z}^s}|a_{f,\nu}(j)|^2\\ &= \sum_{j\in\mathbb{Z}^s} 2^{\nu s}\left|\int_{\mathbb{R}^s} f(x)\overline{K(2^\nu x-j)}\,dx\right|^2\\ &\le \sum_{j\in\mathbb{Z}^s} 2^{\nu s}\|f\|_2^2\|K(2^\nu\cdot -j)\|^2_{2,[-R..R]^s}\\ &\le \|f\|_2^2\|K\|^2_{2,\cup_j(j+2^\nu[-R..R]^s)},\end{aligned}$$

and this shows that $\|P_\nu f\|_2\to 0$ as $\nu\to-\infty$.

Item (iii) follows from the definition of V_ν. Finally, for (iv), since $\{K(\cdot - j)\}_{j\in\mathbb{Z}^s}$ is an orthonormal basis for V_0, the mapping $a\mapsto K{*}a$ is an isomorphism that commutes with τ_j, $j\in\mathbb{Z}^s$. □

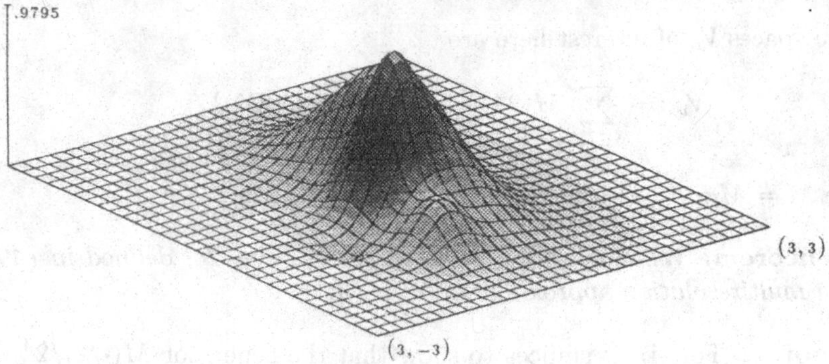

Original function

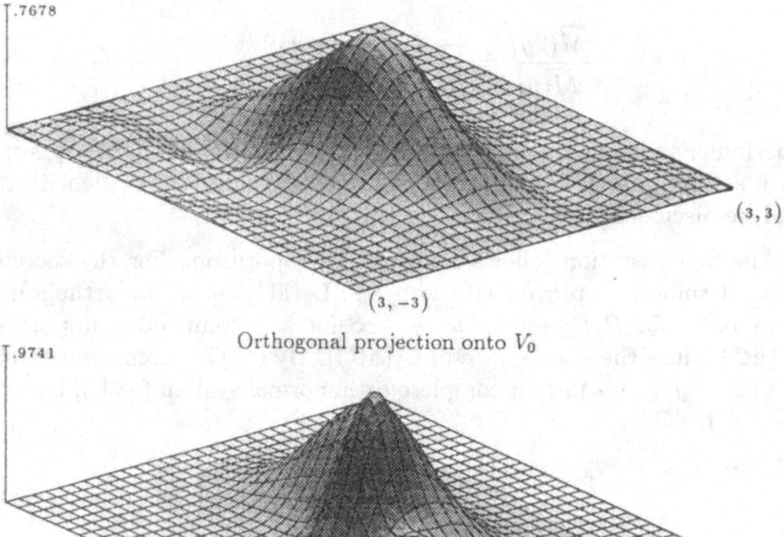

Orthogonal projection onto V_0

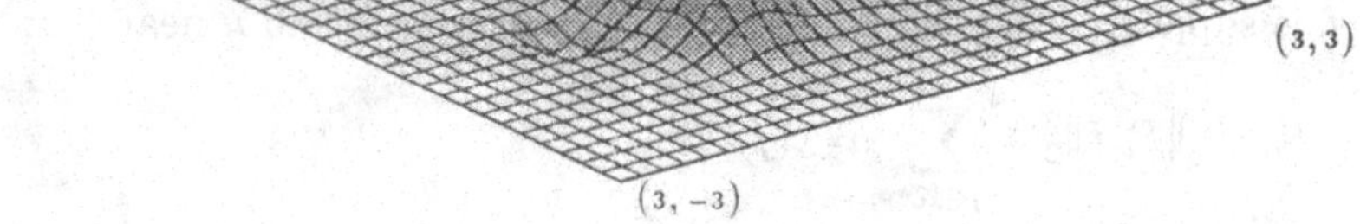

Orthogonal projection onto V_2

(51)Figure. The orthogonal projection (best approximation) of a function onto the spaces V_0 and V_2 generated by the box spline $M_{2,2,2}$.

Example. In (51)Figure, the best approximation from $V_0 = S_2(M)$ for $M = M_{2,2,2}$ is given for a function supported in $[-3\,..\,3]^2$. Also shown is the best approximation from V_2 which is almost indistinguishable graphically from the original function. The best approximation is given by the K-cardinal splines for the function (cf. (48)Corollary). ◻

For the remainder of this discussion we assume that $M = M_\Xi$ with Ξ unimodular. In this case, by (45)Theorem, V_ν can also be described as

$$V_\nu = \{\sum_{j\in\mathbb{Z}^s} K(2^\nu \cdot -j)a(j) : a \in \ell_2(\mathbb{Z}^s)\}.$$

The **refinement equation** for the function K linking the generators of V_0 and V_{-1} is

$$K(\cdot/2)/2^s = K{*}a_H \tag{52}$$

with the coefficient sequence a_H given from the conjugate Fourier series of the function

$$\begin{aligned} H(y) := \frac{\widehat{K}(2y)}{\widehat{K}(y)} &= \frac{\widehat{M}(2y)([\widehat{M},\widehat{M}](y))^{1/2}}{\widehat{M}(y)([\widehat{M},\widehat{M}](2y))^{1/2}} \\ &= \left(\frac{[\widehat{M},\widehat{M}](y)}{[\widehat{M},\widehat{M}](2y)}\right)^{1/2} \prod_{\xi\in\Xi} \frac{1+\exp(-i\xi y)}{2}; \end{aligned} \tag{53}$$

that is,
(54)

$$a_H(j) := \frac{1}{(2\pi)^s}\int_{[-\pi..\pi]^s} \left(\frac{[\widehat{M},\widehat{M}](y)}{[\widehat{M},\widehat{M}](2y)}\right)^{1/2} \prod_{\xi\in\Xi} \frac{1+\exp(-i\xi y)}{2} \exp(ijy)\, dy$$

for $j \in \mathbb{Z}^s$. By the positivity and symmetry of $[\widehat{M},\widehat{M}]$, these coefficients are real-valued and have exponential decay.

The family $\{K(2^\nu \cdot -j)\}_{j\in\mathbb{Z}^s}$ is an orthogonal basis for V_ν, but this orthogonality does not persist between levels ν in general. A desirable goal is to find a finite collection $\{K_\kappa\}$ of functions whose dyadic dilates and dyadic translates

$$\{K_\kappa(2^\nu \cdot -j) : \nu \in \mathbb{Z},\ j \in \mathbb{Z}^s\} \tag{55}$$

form a complete orthogonal system in $\mathbf{L}_2(\mathbb{R}^s)$. Such a collection $\{K_\kappa\}$ is called a **wavelet set**.

We begin by looking for the orthogonal complement, $V_{\nu-1}^\perp$, of $V_{\nu-1}$ in V_ν. More precisely, we search for an orthogonal decomposition of V_0 into 2^s spaces, one of which being V_{-1}, where each space is generated by the (orthogonal) dyadic translates of a single function. For reasons that will become evident later on, the 2^s spaces will be indexed by

$$\mathbb{Z}_2^s := \mathbb{Z}^s/2\mathbb{Z}^s$$

(which can be identified with the vertices of the unit cube $\{0,1\}^s$). Assume for the moment that

$$V_0 = V_{-1} \oplus V_{-1}^{\perp} = V_{-1} \oplus \bigoplus_{\kappa \in \mathbb{Z}_2^s \backslash 0} O_{-1,\kappa}, \tag{56}$$

with the spaces $O_{\nu,\kappa}$ generated by the functions $2^{\nu s} K_\kappa(2^\nu \cdot)$:

$$O_{\nu,\kappa} := \{\sum_{j \in \mathbb{Z}^s} 2^{\nu s} K_\kappa(2^\nu \cdot - j) b(j) : b \in \ell_2(\mathbb{Z}^s)\}, \qquad \nu \in \mathbb{Z}, \quad \kappa \in \mathbb{Z}_2^s \backslash 0.$$

Then, by properties (i,ii) of a multiresolution approximation, we obtain an orthogonal decomposition of $\mathbf{L}_2(\mathbb{R}^s)$:

$$\left(\cup_{\nu \in \mathbb{Z}} V_\nu\right)^- = \bigoplus_{\nu \in \mathbb{Z}} \bigoplus_{\kappa \in \mathbb{Z}_2^s \backslash 0} O_{\nu,\kappa}. \tag{57}$$

Therefore, the problem is to find a set of 2^s pairwise orthogonal functions K_κ, $\kappa \in \mathbb{Z}_2^s$, with $K_0 = K$ and such that

(a) each K_κ has orthogonal shifts;

(b) $K_\kappa(\cdot/2)/2^s \in V_0$; and

(c) for any $a \in \ell_2(\mathbb{Z}^s)$, there exist sequences $b_\kappa \in \ell_2(\mathbb{Z}^s)$ for which

$$\sum_{j \in \mathbb{Z}^s} K(\cdot - j) a(j) = \sum_{\kappa \in \mathbb{Z}_2^s} \sum_{j \in \mathbb{Z}^s} 2^{-s} K_\kappa(\cdot/2 - j) b_\kappa(j). \tag{58}$$

The Requirement (b) is equivalent to the existence of 2π-periodic functions H_κ in $\mathbf{L}_2([0\,..\,2\pi]^s)$ such that

$$\widehat{K}_\kappa(2y) = (K_\kappa(\cdot/2)/2^s)\widehat{\ }(y) = H_\kappa(y)\widehat{K}(y), \quad \kappa \in \mathbb{Z}_2^s. \tag{59}$$

Since $\widehat{K}$ satisfies

$$\sum_{j \in \mathbb{Z}^s} |\widehat{K}(\cdot + 2\pi j)|^2 = 1, \tag{60}$$

the orthonormality of functions K_κ satisfying (59), i.e., Requirement (a) in conjunction with (b), is equivalent to having, for $\kappa, \upsilon \in \mathbb{Z}_2^s$ and arbitrary

$j \in \mathbb{Z}^s$,

(61)

$$
\begin{aligned}
\delta(j)\delta(\kappa - v)
&= \int_{\mathbb{R}^s} K_\kappa(x-j)\overline{K_v(x)}\,dx \\
&= \frac{1}{\pi^s}\int_{\mathbb{R}^s} \widehat{K}_\kappa(2y)\overline{\widehat{K}_v(2y)}\exp(-2ijy)\,dy \\
&= \frac{1}{\pi^s}\int_{\mathbb{R}^s} H_\kappa(y)\overline{H_v(y)}|\widehat{K}(y)|^2\exp(-2ijy)\,dy \\
&= \frac{1}{\pi^s}\int_{[0..2\pi]^s} H_\kappa(y)\overline{H_v(y)}\sum_{j\in\mathbb{Z}^s}|\widehat{K}(y+2\pi j)|^2\exp(-2ijy)\,dy \\
&= \frac{1}{\pi^s}\int_{[0..\pi]^s}\sum_{\kappa\in\mathbb{Z}_2^s} H_\kappa(y+\pi\mu)\overline{H_v(y+\pi\mu)}\exp(-2ijy)\,dy \\
&= \frac{1}{(2\pi)^s}\int_{[0..2\pi]^s}\sum_{\kappa\in\mathbb{Z}_2^s} H_\kappa(y/2+\pi\mu)\overline{H_v(y/2+\pi\mu)}\exp(-ijy)\,dy.
\end{aligned}
$$

Equivalently,

(62)
$$
\begin{aligned}
&\sum_{\mu\in\mathbb{Z}_2^s} |H_\kappa(\cdot+\pi\mu)|^2 = 1, \qquad \kappa\in\mathbb{Z}_2^s, \quad \text{and} \\
&\sum_{\mu\in\mathbb{Z}_2^s} H_\kappa(\cdot+\pi\mu)\overline{H_v(\cdot+\pi\mu)} = 0, \qquad \kappa\neq v.
\end{aligned}
$$

As for Requirement (c), we take the Fourier transform of (58) and use (59), and find that

(63)
$$
A(y)\widehat{K}(y) = \sum_{\kappa\in\mathbb{Z}_2^s} B_\kappa(y)H_\kappa(y)\widehat{K}(y), \quad \text{or} \quad A(y) = \sum_{\kappa\in\mathbb{Z}_2^s} B_\kappa(y)H_\kappa(y),
$$

where

$$
A(y) := \sum_{j\in\mathbb{Z}^s} a(j)\exp(-ijy) \text{ and } B_\kappa(y) := \sum_{j\in\mathbb{Z}^s} b_\kappa(j)\exp(-i2jy), \quad \kappa\in\mathbb{Z}_2^s.
$$

In particular, the B_κ are π-periodic in each variable, and this allows the expansion of (63) into 2^s equations in the unknowns B_κ:

(64)
$$
A(y+\pi\mu) = \sum_{\kappa\in\mathbb{Z}_2^s} H_\kappa(y+\pi\mu)\,B_\kappa(y), \quad \forall \mu\in\mathbb{Z}_2^s.
$$

We conclude the following. Suppose that, for the collection $(H_\kappa)_{\kappa\in\mathbb{Z}_2^s}$ of functions in $\mathbf{L}_\infty([0..2\pi]^s)$, and for each $y\in\mathbb{R}^s$, the matrix

(65)
$$
W(y) := \big(H_\kappa(y+\pi\mu)\big)_{\mu,\kappa\in\mathbb{Z}_2^s}
$$

is unitary, and *define* K_κ by (59). Then, Requirement (a) is satisfied, and (62) follows, therefore Requirement (b) is satisfied. Finally, the functions B_κ given by

$$\big(B_\kappa(y)\big)_{\kappa\in\mathbb{Z}_2^s} := \overline{W(y)}^T \big(A(y+\pi\mu)\big)_{\mu\in\mathbb{Z}_2^s}, \quad y \in \mathbb{R}^s,$$

are π-periodic and in $\mathbf{L}_2([0\,..\,\pi]^s)$, and satisfy (64), therefore Requirement (c) is satisfied.

Thus, the problem is reduced to the construction of $\mathbf{L}_\infty$-functions H_κ for which the matrices in (65) are unitary. Since we want $K_0 = K$, we must have $H_0 = H$. This choice is consistent with having $W(y)$ unitary for every y, since $H_0 = H$ satisfies the first equation of (62): We use (60) twice and the periodicity of H to find

$$\begin{aligned} 1 &= \sum_{j\in\mathbb{Z}^s} |\widehat{K}(2y+2\pi j)|^2 = \sum_{j\in\mathbb{Z}^s} |H(y+\pi j)|^2 |\widehat{K}(y+\pi j)|^2 \\ &= \sum_{\mu\in\mathbb{Z}_2^s}\sum_{j\in\mathbb{Z}^s} |H(y+\pi\mu+2\pi j)|^2 |\widehat{K}(y+\pi\mu+2\pi j)|^2 \\ &= \sum_{\mu\in\mathbb{Z}_2^s} |H(y+\pi\mu)|^2. \end{aligned} \tag{66}$$

Note that this holds indeed for every y since H is continuous.

Thus, we can think of our problem as having been given a vector of unit length, namely $\big(H(y+\pi\mu)\big)_{\mu\in\mathbb{Z}_2^s}$, hence suitable as the first column of a unitary matrix, and wanting to construct the remaining columns of that unitary matrix. Unfortunately, this extension cannot be done in an arbitrary way since we also require that the κ-column have the form $\big(H_\kappa(y+\pi\mu)\big)_{\mu\in\mathbb{Z}_2^s}$. For this reason, we now make the 'Ansatz'

$$H_\kappa(y) := 2^{-s/2} \sum_{\upsilon\in\mathbb{Z}_2^s} \exp(-iy\upsilon)\, G_{\upsilon,\kappa}(2y), \qquad \kappa\in\mathbb{Z}_2^s, \tag{67}$$

with $\big(G_{\upsilon,\kappa}(2y)\big)$ a unitary matrix and the $G_{\upsilon,\kappa}$ 2π-periodic, and in $\mathbf{L}_\infty$. This ensures that the H_κ are 2π-periodic and in $\mathbf{L}_\infty$. Further,

$$\begin{aligned} W(y) &= \big(H_\kappa(y+\pi\mu)\big)_{\mu,\kappa\in\mathbb{Z}_2^s} \\ &= \Big(2^{-s/2}\exp(-i(y+\pi\mu)\upsilon)\Big)_{\mu,\upsilon\in\mathbb{Z}_2^s} \big(G_{\upsilon,\kappa}(2y)\big)_{\upsilon,\kappa\in\mathbb{Z}_2^s} \end{aligned}$$

and this is indeed unitary, by inspection.

Now note that (67) and $H_0 = H$ implies

$$G_{\mu,0} = 2^{s/2} \sum_{j\in\mathbb{Z}^s} a_H(2j+\mu)\exp(-ij\cdot), \qquad \mu\in\mathbb{Z}_2^s, \tag{68}$$

hence it remains to construct the unitary matrix $\big(G_{v,\kappa}(y)\big)$ in any manner whatsoever from its first column. Since any entry of a unitary matrix is bounded by 1 in absolute value, the resulting functions $G_{v,\kappa}$ are in $\mathbf{L}_\infty$ if they are measurable.

There are several possible ways to construct a unitary matrix with given first column. An elementary way is to make use of the Householder matrix $\mathcal{H}(w)$, defined for any $w \in \mathbb{C}^d$ by the rule

$$\mathcal{H} : w \mapsto 1\!\!1 - 2([w]\,[\overline{w}]^T/\overline{w}w).$$

The Householder matrix $\mathcal{H}(w)$ is Hermitian and unitary, i.e., $\overline{\mathcal{H}(w)}^T = \mathcal{H}(w) = \mathcal{H}(w)^{-1}$. Moreover, with

$$\sigma(z) := \begin{cases} z/|z|, & \text{if } z \in \mathbb{C}\backslash\{0\}, \\ 1, & \text{if } z = 0, \end{cases}$$

we have that for any unit vector $w \in \mathbb{C}^d$,

$$\mathcal{H}(w + \sigma(w(1))\mathbf{i}_1)w = -\sigma(w(1))\mathbf{i}_1$$

(a fact used extensively in the QR algorithm). Therefore, the matrix

$$\text{(69)} \qquad \mathcal{H}(w + \sigma(w(1))\mathbf{i}_1)\,\mathrm{diag}(-\sigma(w(1)), 1, \ldots, 1)$$

is unitary and has w as its first column.

(70)Theorem. *Let K be the K-cardinal spline for $M = M_\Xi$ with Ξ unimodular and with refinement equation (52) determined by H in (53). Let the functions K_κ be given by $\widehat{K}_\kappa = H_\kappa \widehat{K}$ where H_k are given in (67) with $G_{\mu,0}$ from (68) and for $\kappa \neq 0$*

$$\text{(71)} \qquad G_{v,\kappa} := \begin{cases} -\sigma(G_{0,0})\overline{G_{\kappa,0}}, & \textit{when } v = 0, \\ \delta_{v,\kappa} - G_{v,0}\overline{G_{\kappa,0}}/(1 + |G_{0,0}|), & \textit{otherwise.} \end{cases}$$

Then the functions

$$\{K_\kappa(2^\nu \cdot -j) : j \in \mathbb{Z}^s,\ \nu \in \mathbb{Z},\ \kappa \in \mathbb{Z}_2^s\backslash\{0\}\}$$

form a complete orthogonal set for $\mathbf{L}_2(\mathbb{R}^s)$.

Proof. With $w = (G_{\mu,0})_{\mu \in \mathbb{Z}_2^s}$ and the observation that

$$\overline{(w + \sigma(w(1))\mathbf{i}_1)}(w + \sigma(w(1))\mathbf{i}_1) = 2(1 + |w(1)|),$$

the entries of the matrix (69) are precisely those in (71). Since the $G_{\mu,0}$ are continuous, the remaining $G_{v,\kappa}$ are measurable, hence in $\mathbf{L}_\infty$. ◻

(72)Remark. If $\widehat{M}$ is nonnegative, then so is H. Since $\sum_{\mu\in\mathbb{Z}_2^s}|H(y+\pi\mu)|^2=1$, this implies that $G_{0,0}(y)=2^{-s/2}\sum_{\mu\in\mathbb{Z}_2^s}H(y+\pi\mu)>0$ for all y. Consequently, $\sigma(G_{0,0})=1$ and $|G_{0,0}|=G_{0,0}$. Hence, with (71), we conclude that all the $G_{\upsilon,\kappa}$ have exponentially decaying Fourier coefficients, since this is certainly true for each $G_{\upsilon,0}$. Consequently, the wavelets K_κ constructed here decay exponentially in this case. For example, for unimodular matrices with only even multiplicities, the center c_Ξ is an integer so that $M=M_\Xi^c$ generates V_0 and has $\widehat{M}\geq 0$.

Wavelets in dimensions ≤ 3. In low dimensions ($s\leq 3$), there is a simple construction, which gives the functions K_κ in the form $K{*}a_{H_\kappa}$, with all the sequences a_{H_κ} obtained from the sequence a_H in (54) by translation and change of sign pattern. Moreover, the K_κ all have exponential decay. The construction depends on the center, $c_\Xi=\sum_{\xi\in\Xi}\xi/2$, of the box spline. Define

$$(73)\qquad H_\kappa^*:=\begin{cases}H, & \text{if } 2\kappa c_\Xi \text{ is even;}\\ \overline{H}, & \text{if } 2\kappa c_\Xi \text{ is odd;}\end{cases}\qquad \kappa\in\mathbb{Z}_2^s,$$

where H is given in (53), and choose the functions H_κ to be of the form

$$(74)\qquad H_\kappa=\exp(i\eta(\kappa)\cdot\,)H_\kappa^*(\cdot+\pi\kappa),\qquad \kappa\in\mathbb{Z}_2^s,$$

with $\eta(\kappa)\in\mathbb{Z}_2^s$. We now verify that the resulting matrices $W(y)$ in (65) are unitary, provided the map $\eta:\mathbb{Z}_2^s\to\mathbb{Z}_2^s$ satisfies the condition

$$(75)\qquad \eta(0)=0 \quad\text{and}\quad (\eta(\kappa)+\eta(\upsilon))(\kappa+\upsilon) \quad\text{is odd if}\quad \kappa\neq\upsilon.$$

By (66), the first condition in (62) is met for every $\kappa\in\mathbb{Z}_2^s$. For $\kappa\neq\upsilon$, we consider the terms of the second relation in (62) corresponding to μ and $\mu+\kappa+\upsilon$ (these are separate indices since $\kappa+\upsilon\neq 0$ in $\mathbb{Z}_2^s$). The first term is

$$\begin{aligned}&H_\kappa(\cdot+\pi\mu)\overline{H_\upsilon(\cdot+\pi\mu)}\\ &=\exp(i(\eta(\kappa)-\eta(\upsilon))\cdot\,)(-1)^{(\eta(\kappa)-\eta(\upsilon))\mu}H_\kappa^*(\cdot+\pi\kappa+\pi\mu)\overline{H_\upsilon^*(\cdot+\pi\upsilon+\pi\mu)},\end{aligned}$$

while by the 2π-periodicity, the second term is

$$\exp(i(\eta(\kappa)-\eta(\upsilon))\cdot\,)(-1)^{(\eta(\kappa)-\eta(\upsilon))(\mu+\kappa+\upsilon)}H_\kappa^*(\cdot+\pi\upsilon+\pi\mu)\overline{H_\upsilon^*(\cdot+\pi\kappa+\pi\mu)}.$$

We claim that these two terms will be equal except of opposite sign by (75) provided

$$H_\kappa^*(\cdot+\pi\kappa+\pi\mu)\overline{H_\upsilon^*(\cdot+\pi\upsilon+\pi\mu)}=H_\kappa^*(\cdot+\pi\upsilon+\pi\mu)\overline{H_\upsilon^*(\cdot+\pi\kappa+\pi\mu)}.$$

The last relation follows directly from the definition of H_κ^* if one of $2\kappa c_\Xi$ and $2v c_\Xi$ is even and the other odd, for then H_κ^* and $\overline{H_v^*}$ are either both H or both $\overline{H}$. In the opposite case, when one of H_κ^* and $\overline{H_v^*}$ is H and the other is $\overline{H}$, the proof requires the observation, obtainable from (53), that

$$\overline{H(y+\pi\mu)} = \exp(i2yc_\Xi)(-1)^{2\mu c_\Xi} H(y+\pi\mu). \tag{76}$$

Then if, say, both $2\kappa c_\Xi$ and $2v c_\Xi$ are odd, we have

$$\begin{aligned} H_\kappa^*(\cdot + \pi\kappa + \pi\mu)&\overline{H_v^*(\cdot + \pi v + \pi\mu)} \\ &= \exp(i2c_\Xi \cdot)(-1)^{2(\mu+\kappa)c_\Xi} H(\cdot + \pi\mu + \pi\kappa)H(\cdot + \pi\mu + \pi v) \\ &= \exp(i2c_\Xi \cdot)(-1)^{2(\mu+v)c_\Xi} H(\cdot + \pi\mu + \pi v)H(\cdot + \pi\mu + \pi\kappa) \\ &= H_\kappa^*(\cdot + \pi v + \pi\mu)\overline{H_v^*(\cdot + \pi\kappa + \pi\mu)}. \end{aligned}$$

When the H_κ are chosen by (74), their Fourier coefficient sequences exhibit the same exponential decay as the coefficients for H. Moreover, since

$$\widehat{K}_\kappa(2\cdot) = H_\kappa \widehat{M}/[\widehat{M}, \widehat{M}]^{1/2},$$

and the Fourier coefficients of each function $H_\kappa/[\widehat{M}, \widehat{M}]^{1/2}$ has exponential decay, the functions K_κ must have exponential decay (cf. (45)Theorem).

We now exhibit 1-1 mappings $\eta : \mathbb{Z}_2^s \to \mathbb{Z}_2^s$ that satisfy (75) for $s = 1, 2, 3$. For $s = 1$, $\eta(0) = 0$ and $\eta(1) = 1$. When $s = 2$, one choice of the mapping η is

$$\begin{matrix} (0,0) \mapsto (0,0) & (0,1) \mapsto (0,1) \\ (1,0) \mapsto (1,1) & (1,1) \mapsto (1,0). \end{matrix} \tag{77}$$

Finally, a suitable mapping η in the case of $s = 3$ is

$$\begin{aligned} &(0,0,0) \mapsto (0,0,0),\ (1,0,0) \mapsto (1,1,0),\ (0,1,0) \mapsto (0,1,1), \\ &(1,1,0) \mapsto (1,0,0),\ (0,0,1) \mapsto (1,0,1),\ (1,0,1) \mapsto (0,0,1), \\ &(0,1,1) \mapsto (0,1,0),\ (1,1,1) \mapsto (1,1,1). \end{aligned}$$

In other words, η is self-inverse, fixes $(0,0,0)$ and $(1,1,1)$, and maps $\mathbf{i}_j$ to $\mathbf{i}_j + \mathbf{i}_{j+1}$.

The form (74) of the functions H_κ makes it particularly easy to determine the functions K_κ from K when the sequence a_H from (52) is known.

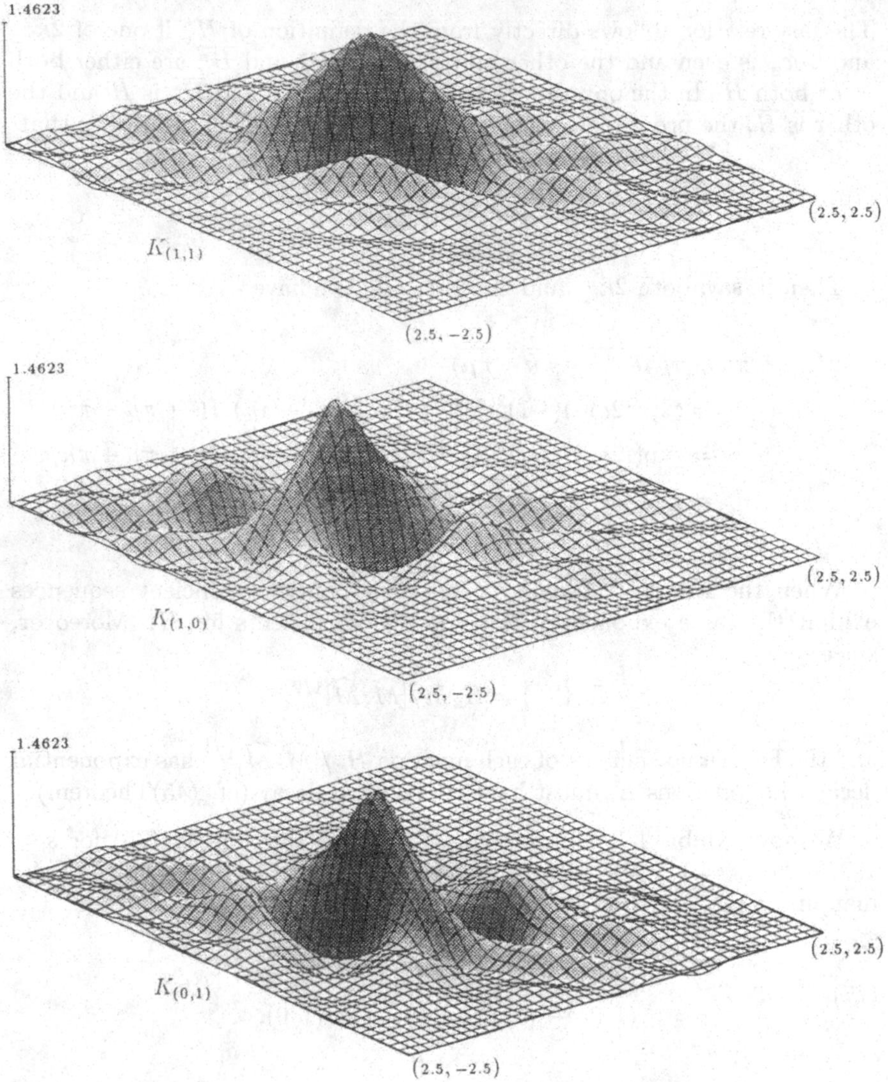

(78)Figure. The wavelets for $M^c_{2,2,2}$ on $[-2.5\,..\,2.5]^2$.

(79)Theorem. *Let K be the K-cardinal spline for $M = M_\Xi$ with Ξ unimodular and with refinement equation (52). If the dimension $s \le 3$, then the functions*

$$K_\kappa = K{*}a_{H_\kappa},$$

where

$$a_{H_\kappa}(j) = (-1)^{j\kappa+1} a_H((-1)^{2\kappa c_\Xi}(j + \eta(\kappa))),$$

are a family of wavelets, i.e.,

$$\left\{K_\kappa(2^\nu \cdot -j) : j \in \mathbb{Z}^s,\ \nu \in \mathbb{Z},\ \kappa \in \mathbb{Z}_2^s\backslash\{0\}\right\}$$

forms a complete orthogonal set in $\mathbf{L}_2(\mathbb{R}^s)$. *Moreover,*

$$(80) \qquad |K_\kappa(x)| \le \text{const}\exp(-|x|/\text{const}), \qquad \kappa \in \mathbb{Z}_2^s,$$

for some positive constants.

Proof. This is an immediate consequence of the construction above. The form of the coefficients is derived from (74) and the fact that a_H is the sequence of coefficients for H. □

(81)Remark. It is impossible to find the mapping η for $s \ge 4$, for the following reason. For any such η and for any real unit vector $(x_v)_{v\in\mathbb{Z}_2^s}$, the matrix

$$\left((-1)^{\eta(\mu)v} x_{v-\mu}\right)_{\mu,v\in\mathbb{Z}_2^s}$$

is real, unitary, and has (x_v) as its first row. However, such matrices are known to exist only for $s = 0, 1, 2, 3$.

Notes. The *striking result* mentioned at the outset is Schoenberg's; it is the major result in [Schoenberg'74a]. The specific generalization of Schoenberg's result to box splines is (29)Theorem, taken, along with its proof, from [de Boor, Höllig, Riemenschneider'86b], as is the basic (4)Proposition concerning fundamental domains. The discussion in (15)Example of fundamental domains for box splines on a two-direction mesh and their associated tilings (along with the figures) comes from [de Boor, Höllig'91]. The details for (22)Example are taken from [de Boor, Höllig, Riemenschneider'85b]. Related papers are: [de Boor, Höllig, Riemenschneider'85a,'86b,'87], [Höllig, Marsden, Riemenschneider'89].

A good reference for the basic facts on Fourier analysis is [Rudin'73]; e.g., the Paley-Wiener theorem is Theorem 7.23 there. The basic facts about Householder matrices can be found in any good book on Numerical Analysis or Numerical Linear Algebra, e.g., [Ciarlet'88].

A wealth of information about *cardinal series* can be found in the article of Higgins [Higgins'85]. It was there that we learned of Gosselin's article [Gosselin'63], whose definition of K-cardinal series we adopted. This approach seems to fit perfectly with cardinal splines and cardinal interpolation.

Multiresolution analysis was introduced in [Meyer'90] and [Mallat'89a] in the study of *wavelet decompositions*. Extensions and refinements of this theory that influenced the presentation and would generalize it are provided in [Jia, Micchelli'92a], [de Boor, DeVore, Ron'93], and [Jia, Shen'9x].

The construction in dimensions ≤ 3 for (79)Theorem by [Riemenschneider, Shen'91] predates (70)Theorem, but a nonconstructive existence theorem for multivariate wavelets appeared already in [Gröchenig'87] and was included in [Meyer'90]. Also, B-splines had been used in univariate constructions of wavelets in [Battle'87] and [Meyer'90]. (81)Remark has long been known to combinatorists and is credited to [Hurwitz'23].

The problem of extending a single row to a unitary matrix as a means to constructing wavelets was discussed in [Meyer'90] and independently was one of the main themes in [Jia, Micchelli'91a,'92a]. The latter also studied the construction of prewavelets which satisfy the orthogonality relations

$$K(2^\nu \cdot - j) \perp K(\cdot - k) \quad \forall j, k \in \mathbb{Z}^s, \quad \text{if } \nu \neq 0.$$

(In other words, they have orthogonality between dyadic levels, but the shifts within a given level are not orthogonal.) Prewavelets are interesting because it is easier to construct compactly supported ones. Explicit compactly supported box spline prewavelets were constructed in dimensions $s = 2, 3$ by [Riemenschneider, Shen'92], [Chui, Stöckler, Ward'92b], and [Jia, Micchelli'91a]. For $s = 1$, the relevant papers are [Chui, Wang'92a,'92b]. Other papers giving constructions of wavelets and prewavelets in higher dimensions using box splines are [Lorentz, Madych'92a], [Stöckler'92], [de Boor, DeVore, Ron'93], and [Jia, Shen'9x].

The Householder matrix was used in prewavelet and wavelet constructions by [Jia, Micchelli'92a] and [Jia, Shen'9x], the construction for (70) Theorem is taken from the latter paper. The fact that the possibly discontinuous function $\sigma(w)$ appears there hints at one of the difficulties in extending the given column to a unitary matrix to gain wavelets with nice decay properties: the nonexistence of sufficiently many independent tangent vector fields on spheres for $s > 3$ ([Adams'62]). Using specific properties of the box splines, [Jia, Shen'9x] give an explicit construction of exponentially decaying wavelets in any dimension from box splines with unimodular direction matrices.

VI

Discrete box splines & linear diophantine equations

In this chapter, we give the basic properties of the discrete box spline, in close analogy to those of the (continuous) box spline discussed in Chapter I. This includes facts about the structure of the discrete box spline, as well as the characterization of global and local linear independence. This gives us a chance to stress the perhaps surprising connection between box splines and linear diophantine equations.

Definition. The **discrete box spline** $b = b^h = b^h_\Xi = b^h(\cdot, \Xi)$ associated with the matrix $\Xi \in \mathbb{Z}^{s \times n}$ and $h \in 1/\mathbb{N} := \{1/j : j \in \mathbb{N}\}$ is, by definition, the distribution given by the rule

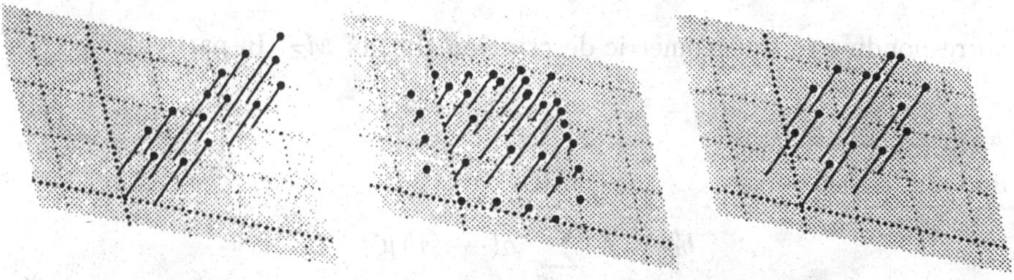

(1)Figure. Discrete box-splines for $h = 1/2$ and $\Xi =$

$$\begin{bmatrix}1 & 0 & 3 & 1\\ 0 & 1 & 1 & 3\end{bmatrix}, \qquad \begin{bmatrix}1 & -1 & 1 & 1 & 1 & 0 & 0 & 0\\ 1 & 1 & 0 & 0 & 0 & 1 & 1 & 1\end{bmatrix}, \qquad \begin{bmatrix}1 & -1 & 1 & 2\\ 1 & 1 & 2 & 1\end{bmatrix}.$$

(2)
$$b^h : C(\mathbb{R}^s) \to \mathbb{R} : \varphi \mapsto \langle b^h, \varphi\rangle := h^n \sum_{\alpha\in\square_h} \varphi(\Xi\alpha),$$

with
$$\square_h := \square \cap h\mathbb{Z}^n = \{0, h, \ldots, 1-h\}^n.$$

Since
$$\lim_{h\to 0} h^n \sum_{\alpha\in\square_h} \varphi(\Xi\alpha) = \int_\square \varphi(\Xi t)\, dt,$$

$b^h = b^h_\Xi$ converges to $M = M_\Xi$ as $h \to 0$, thus justifying the name.

Geometric description. It is convenient to visualize b^h also as a mesh function, i.e.,
$$\langle b^h, \varphi\rangle =: \sum_{k\in h\mathbb{Z}^s} b^h(k)\varphi(k).$$

Then
$$b^h = \sum_{k\in h\mathbb{Z}^s} b^h(k)\delta(\cdot - k),$$

with the distribution $\delta : \varphi \mapsto \varphi(0)$ taken also as a mesh function, viz. the mesh function (on whatever convenient mesh) which takes the value 1 at zero and vanishes everywhere else:
$$\delta(j) = \begin{cases} 1, & \text{if } j = 0; \\ 0, & \text{otherwise.} \end{cases}$$

By considering in (2) any particular φ which is 1 near k and zero outside a small ball around k, we obtain the explicit formula

(3)
$$b^h_\Xi(k) = h^n \#((\Xi^{-1}k)\cap\square_h) = h^n\#\{\alpha \in \square_h : \Xi\alpha = k\} = h^n \sum_{\alpha\in\square_h} \delta(k-\Xi\alpha),$$

corresponding to the geometric description (I.3) of M_Ξ. In particular,
$$b^h_{[]} = \delta,$$

and
$$b^h_{[\zeta]} = h \sum_{j=0}^{1/h-1} \delta(\cdot - (jh)\zeta).$$

Support. The support of b^h_Ξ is the set $\Xi\square_h$, and b^h is positive on its support. But simple examples (see, e.g., (1)Figure above, (8)Figure below, or (VII.13)Figure in the next chapter) show that the support of b^h need not be convex. (Here, we call a subset G of $h\mathbb{Z}^s$ **convex** if it is the intersection of its convex hull (as a subset of $\mathbb{R}^s$) with $h\mathbb{Z}^s$.)

Convolution. In analogy to (I.5),

$$\sum_{h\mathbb{Z}^s} b_\Xi^h := \sum_{k\in h\mathbb{Z}^s} b^h(k) = \langle b^h, 1\rangle = h^n \sum_{\alpha\in\Box_h} 1 = 1,$$

while the discrete version of the convolution identity (I.18) is

$$(4)\qquad b_Z^h * b_Y^h = \sum_{k\in h\mathbb{Z}^s} b_Z^h(\cdot - k) b_Y^h(k) = b_{Z\cup Y}^h.$$

Here is a proof of (4) based on (3) (in which we carry over from $\Box$ the convenient agreement that the dimension of $\Box_h$ is determined by the context):

$$\begin{aligned} b_Z^h * b_Y^h &= \sum_{\ell\in h\mathbb{Z}^s} \sum_{\alpha\in\Box_h} \sum_{\beta\in\Box_h} h^{\#Z}\delta(\cdot - \ell - Z\alpha)\, h^{\#Y}\delta(\ell - Y\beta) \\ &= h^{\#(Z\cup Y)} \sum_{(\alpha,\beta)\in\Box_h} \delta(\cdot - (Z\cup Y)(\alpha,\beta)) = b_{Z\cup Y}^h. \end{aligned}$$

Construction. The identity (4) implies that

$$(5)\qquad b_{Z\cup\zeta}^h = h \sum_{\alpha\in\Box_h} b_Z^h(\cdot - \zeta\alpha) = h \sum_{j=0}^{1/h-1} b_Z^h(\cdot - jh\zeta).$$

Thus, starting with

$$b_{[]}^h = \delta,$$

(5) provides a simple algorithm for computing b^h.

(6)Example. As an illustration, we compute the discrete box spline corresponding to $\Xi = \begin{bmatrix} 1 & 0 & 1 & -1 \\ 0 & 1 & 1 & 1 \end{bmatrix}$ and $h = 1/2$, i.e., the discrete ZP element. Starting with

$$b_{[]}^h = \delta,$$

averaging in the direction $\xi = \mathbf{i}_1$ yields

$$2b_{[\mathbf{i}_1]}^h(k) = \delta(k) + \delta(k - \mathbf{i}_1/2),$$

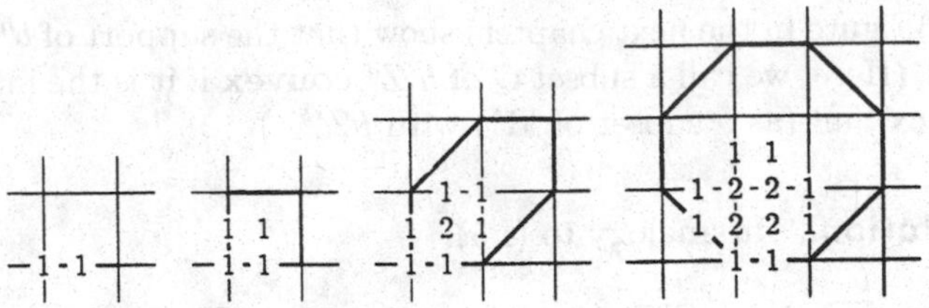

(7)Figure. The discrete box-spline $b^h(\cdot, \left[\begin{smallmatrix}1&0&1&-1\\0&1&1&1\end{smallmatrix}\right])$ for $h = 1/2$ is built up by repeated averaging, from δ through the sequence $b^h(\cdot, \left[\begin{smallmatrix}1\\0\end{smallmatrix}\right])$, $b^h(\cdot, \left[\begin{smallmatrix}1&0\\0&1\end{smallmatrix}\right])$, $b^h(\cdot, \left[\begin{smallmatrix}1&0&1\\0&1&1\end{smallmatrix}\right])$. The scaled mesh functions $b^h_Z/h^{\#Z}$ are shown.

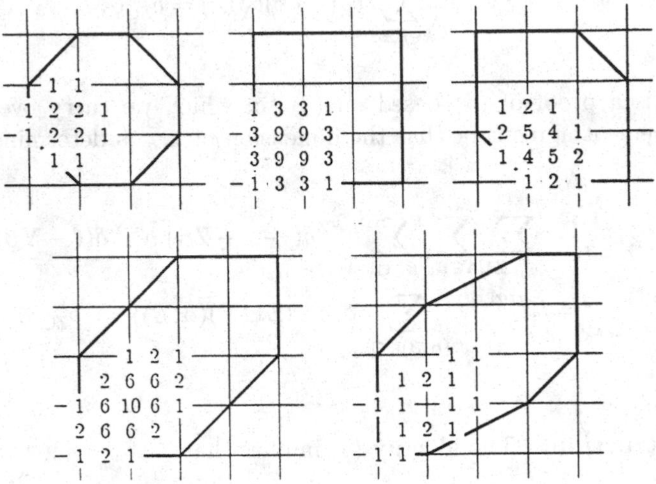

(8)Figure. The mesh functions $b^h_\Xi/h^{\#\Xi}$ for $h = 1/2$ and $\Xi =$

$$\left[\begin{smallmatrix}1&0&1&1\\0&1&1&-1\end{smallmatrix}\right], \left[\begin{smallmatrix}1&1&1&0&0&0\\0&0&0&1&1&1\end{smallmatrix}\right], \left[\begin{smallmatrix}1&1&0&0&-1\\0&0&1&1&1\end{smallmatrix}\right], \left[\begin{smallmatrix}1&1&0&0&1&1\\0&0&1&1&1&1\end{smallmatrix}\right], \left[\begin{smallmatrix}1&2&1&0\\0&1&1&2\end{smallmatrix}\right].$$

i.e., $b^h_{[\mathbf{i}_1]}(k)$ assumes the value $1/2$ for $k \in \{(0,0), (1/2,0)\}$ and vanishes for all other $k \in h\mathbb{Z}^2$. (7)Figure records this result together with those of the subsequent averaging in the directions $\mathbf{i}_2$, $\mathbf{i}_1 + \mathbf{i}_2$ and $-\mathbf{i}_1 + \mathbf{i}_2$.

Fourier transform. We compute from (2) that

$$\sum_{k \in h\mathbb{Z}^s} b^h(k) \exp(-iyk) = \widehat{b^h}(y) = h^n \sum_{\alpha \in \mathsf{L}_h} \exp(-iy\Xi\alpha)$$

$$= h^n \prod_{\xi \in \Xi} \frac{1 - \exp(-iy\xi)}{1 - \exp(-ihy\xi)} = \widehat{M}(y)/\widehat{M}(hy),$$

i.e.,

$$\widehat{b^h} = \widehat{M}/\widehat{M}(h\cdot), \qquad (9)$$

using the fact that, by (I.17), $\widehat{M}(y) = \widehat{M_\Xi}(y) = \prod_{\xi\in\Xi} \frac{1-\exp(-iy\xi)}{iy\xi}$ is the Fourier transform for $M = M_\Xi$.

Consequently,

$$\widehat{M}(y) = \sum_{k\in h\mathbb{Z}^s} b^h(k)\exp(-iyk)\widehat{M}(hy),$$

and therefore

$$M = M^h * b^h = \sum_{k\in h\mathbb{Z}^s} M^h(\cdot - k)b^h(k), \tag{10}$$

with

$$M^h := M(\cdot/h)/h^s.$$

This identity is at the basis of the subdivision algorithm for the evaluation of the box spline series $M*a = \sum_{j\in\mathbb{Z}^s} M(\cdot - j)a(j)$ which is the subject of the next chapter. For that discussion, we also need the identity

$$b^{hh'} = b^h * (\sigma_h b^{h'}) = \sum_k b^h(\cdot - k)b^{h'}(k/h), \tag{11}$$

which follows from (9) with the observation that

$$\frac{\widehat{M}}{\widehat{M}(hh'\cdot)} = \frac{\widehat{M}}{\widehat{M}(h\cdot)}\frac{\widehat{M}(h\cdot)}{\widehat{M}(hh'\cdot)}.$$

For $h = 1/2$, (9) simplifies to

$$\widehat{b^{1/2}} = \prod_{\xi\in\Xi} \frac{1+\exp(-\cdot\, i\xi/2)}{2}. \tag{12}$$

Differentiation. Since

$$\frac{\widehat{M_\Xi}}{\widehat{M_\Xi}(h\cdot)}(1-\exp(-\cdot\, ih\xi)) = h\frac{\widehat{M_{\Xi\backslash\xi}}}{\widehat{M_{\Xi\backslash\xi}}(h\cdot)}(1-\exp(-\cdot\, i\xi))$$

for any $\xi \in \Xi$, we conclude from (9) that

$$\nabla_{h\xi} b^h_\Xi = h\nabla_\xi b^h_{\Xi\backslash\xi}.$$

This proves the following discrete analog

$$\nabla_{hZ} b^h_\Xi = h^{\#Z}\nabla_Z b^h_{\Xi\backslash Z} \tag{13}$$

of the differentiation formula (I.30), valid for arbitrary $Z \subseteq \Xi$. In particular,

$$\nabla_{h\Xi}\, b^h_\Xi \;=\; h^n \nabla_\Xi \delta,$$

which corresponds to (I.30) with $Z = \Xi$, i.e., to the identity

$$D_\Xi M_\Xi = \nabla_\Xi \delta.$$

Annihilation. The differentiation formula (13) implies that

$$\text{(14)} \qquad \forall\, \{Z \in \mathcal{A}(\Xi)\} \quad \nabla_{hZ}\, b^h = 0 \quad \textit{on } \mathbb{R}^s \backslash \Gamma(\Xi),$$

i.e., with b^h taken as a function on $\mathbb{R}^s$ whose support happens to lie in $h\mathbb{Z}^s$.

Indeed, from (13),

$$\nabla_{hZ} b^h_\Xi(x) = h^{\#Z} \nabla_Z b^h_{\Xi\backslash Z}(x),$$

while

$$\nabla_Z b^h_Y(x) = \sum_{V \subseteq Z} (-1)^{\#V} b^h_Y(x - 2c_V),$$

with $2c_V = \sum_{\xi \in V} \xi$, hence

$$\nabla_Z b^h_{\Xi\backslash Z}(x) = 0$$

if $x \notin \operatorname{supp} b^h_{\Xi\backslash Z}(\cdot - 2c_V) = 2c_V + (\Xi\backslash Z)\square_h$ for all $V \subseteq Z$.

Linear independence. The characterization of linear independence of the shifts of a discrete box spline is richer than that for the box spline itself. It has strong interactions with linear diophantine equations. The theory can be made to cover the more general situation in which h is not just a scalar, but a vector,

$$h = (h_\xi)_{\xi \in \Xi},$$

with each h_ξ the reciprocal of some natural number, and, correspondingly, the mesh in $\mathbb{R}^s$ of interest is not just a scalar multiple of $\mathbb{Z}^s$, but the lattice $\Xi_h \mathbb{Z}^n$, with

$$\Xi_h := [\ldots, \xi h_\xi, \ldots] =: \Xi N^{-1},$$

i.e.,

$$N := \operatorname{diag}\left((N_\xi)_{\xi\in\Xi}\right) := \operatorname{diag}\left((1/h_\xi)_{\xi\in\Xi}\right) \;\in\; \mathbb{N}^{n\times n}.$$

In this more general setting, the definition for b^h_Ξ becomes

$$\langle b^h, \varphi\rangle := \sum_{\alpha\in\square_h} \varphi(\Xi\alpha)/\det N, \tag{15}$$

with $\square_h$ now denoting the discrete set

$$\square_h := \times_{\xi\in\Xi} \{0, h_\xi, \dots, 1-h_\xi\}.$$

It is instructive to derive (15) from the attempt to write $M = M_\Xi$ as a linear combination of certain translates of the box spline M_{Ξ_h}: For any test function $\varphi \in C(\mathbb{R}^s)$, we compute

$$\begin{aligned}
\int_{\mathbb{R}^s} \varphi(x) M_\Xi(x)dx &= \int_\square \varphi(\Xi t)dt \\
&= \int_{N\square} \varphi(\Xi_h u)du/\det N \\
&= \sum_{j\in\mathbb{Z}^n\cap N\square} \int_{j+\square} \varphi(\Xi_h u)du/\det N \\
&= \sum_{j\in\mathbb{Z}^n\cap N\square} \int_\square \varphi(\Xi_h(u+j))du/\det N \\
&= \sum_{j\in\mathbb{Z}^n\cap N\square} \int_{\mathbb{R}^s} \varphi(x) M_{\Xi_h}(x-\Xi_h j)dx/\det N.
\end{aligned}$$

Therefore

$$M_\Xi = \sum_{j\in\Xi_h\mathbb{Z}^n} M_{\Xi_h}(\cdot - j) b^h(j), \tag{16}$$

with

$$b^h(j) := \#\{k\in\mathbb{Z}^n \cap N\square : \Xi_h k = j\}/\det N = \#\{\alpha\in\square_h : \Xi\alpha = j\}/\det N.$$

We recognize (16) as the convolution product

$$M_\Xi = M_{\Xi_h} * b^h$$

of the function M_{Ξ_h} with the *distribution*

$$b^h = \sum_{j\in\Xi_h\mathbb{Z}^n} b^h(j)\delta(\cdot - j),$$

with δ the Dirac measure centered at 0. Consequently,

$$\widehat{M_\Xi} = \widehat{M_{\Xi_h}} \widehat{b^h}.$$

This implies that

$$\widehat{b^h}(y) = \prod_{\xi \in \Xi} h_\xi \frac{1 - \exp(-iy\xi)}{1 - \exp(-iy\xi h_\xi)}, \tag{17}$$

using the formula (I.17) for the Fourier transform of M_Ξ. In particular, $\widehat{b^h}(y) = 0$ if and only if $y\xi \in 2\pi(\mathbb{Z}\backslash N_\xi \mathbb{Z})$ for some $\xi \in \Xi$.

This is important when studying the linear independence of shifts of b^h, because of the following result, recalled here from (IV.65).

(18)Result. *The shifts of the compactly supported distribution b are linearly independent if and only if, for every $y \in \mathbf{C}^s$, there exists a $j \in \mathbb{Z}^s$ such that*

$$\hat{b}(y + 2\pi j) \neq 0. \tag{19}$$

Thus the shifts of b^h are linearly independent if and only if, for every $y \in \mathbf{C}^s$, there exists some $j \in \mathbb{Z}^s$ such that

$$(y + 2\pi j)\xi \notin 2\pi(\mathbb{Z}\backslash N_\xi \mathbb{Z})$$

for every $\xi \in \Xi$. After dividing by 2π, this condition reads that, for every $y \in \mathbf{C}^s$, there exists some $j \in \mathbb{Z}^s$ such that

$$(y + j)\xi \ \notin \ \mathbb{Z}\backslash N_\xi \mathbb{Z} \tag{20}$$

for every $\xi \in \Xi$. This condition is automatically satisfied for every ξ with $y\xi \notin \mathbb{Z}$. For the others, we must choose j so that $(y + j)\xi \ \in \ N_\xi \mathbb{Z}$. Thus the linear independence of the shifts of b^h is equivalent to the existence of an integer solution $(j, k) \in \mathbb{Z}^s \times \mathbb{Z}^\Theta$ to the linear equations

$$\Theta^T j + N(\Theta)k = \Theta^T y,$$

with $\Theta \subseteq \Xi$ consisting of all $\theta \in \Xi$ with $\theta y \in \mathbb{Z}$, and $N(\Theta)$ the diagonal matrix with diagonal entries N_θ, $\theta \in \Theta$. This is a problem in *linear diophantine equations.* We suspend our discussion of the linear independence of the shifts of b^h in order to recall the needed basic facts from that area.

Linear diophantine equations. Throughout this section, let $A \in \mathbb{Z}^{m \times n}$ with $m \leq n$ be of full rank. We are interested in finding integer solutions to the equation

$$A? = c \tag{21}$$

for given $c \in \mathbb{Z}^m$.

The basic theorem (see (26)Theorem below) concerning the solvability of linear diophantine equations parallels the standard theorem on the solvability of linear equations, in that (21) is solvable if and only if A and $[A, c]$ have the same 'rank', but the notion of 'rank' needed here is a bit more subtle than that of matrix rank.

Throughout, we use the notation

$$p \mid q$$

to indicate that p is a divisor of q, i.e., the integer q is an integer multiple of the integer p. We also use

$$\gcd(\Sigma)$$

for the greatest common divisor of all the entries of Σ, with Σ a set, or a vector, or a matrix.

(22)Smith normal form. *For any $A \in \mathbb{Z}^{m \times n}$, there exist unimodular matrices $V \in \mathbb{Z}^{m \times m}$, $U \in \mathbb{Z}^{n \times n}$ so that $\Sigma := V^{-1}AU^{-1}$ is diagonal, with $\Sigma(i,i) \mid \Sigma(i+1,i+1)$ for all i for which $\Sigma(i,i)$ or $\Sigma(i+1,i+1)$ is nonzero. (In particular, $\Sigma(i,i) = 0$ implies that $\Sigma(i+1,i+1) = 0$.)*

This is proved by induction, using row- and column-elimination. The key lemma needed is the following.

(23)Lemma. *For any $a \in \mathbb{Z}^n$ there exists $X \in \mathbb{Z}^{n \times (n-1)}$ so that* $\det[X, a] = \gcd(a)$.

Proof. By induction on n. Let $a =: (b, c) \in \mathbb{Z}^{n-1} \times \mathbb{Z}$ and make the 'Ansatz'

$$D := [X, a] = \begin{bmatrix} Y & r[b] & [b] \\ 0 & d & c \end{bmatrix},$$

with $Y \in \mathbb{Z}^{(n-1)\times(n-2)}$ chosen (as we can by induction hypothesis) so that $\det[Y, b] = \gcd(b)$, and r and d to be determined. Then

$$\det D = c \det[Y, rb] - d \det[Y, b] = (cr - d) \det[Y, b],$$

hence we are done provided we can choose r and d so that $D \in \mathbb{Z}^{n \times n}$ and $cr - d = \gcd(a)/\gcd(b)$, i.e.,

$$\gcd(b)cr - \gcd(b)d = \gcd(a).$$

Since $\gcd(a) = \gcd(\gcd(b), c)$, there are integers μ and ν so that

$$\mu \gcd(b) + \nu c = \gcd(a).$$

Hence the choice $r := \nu/\gcd(b), d := -\mu$ will do the job since, in particular, $rb = \nu b/\gcd(b) \in \mathbb{Z}^{n-1}$. □

Proof of Smith normal form. We conclude from the Lemma, with the aid of permutations and transpositions, that, for any choice of row or column, we can find some $Y \in \mathbb{Z}^{n\times n}$ having a given $a \in \mathbb{Z}^n$ in that row or column and having $\det Y = \gcd(a)$.

In particular, if $\alpha := \gcd(A(:,1))$ for the given $A \in \mathbb{Z}^{m\times n}$, and therefore $\alpha = [a]^T A(:,1)$ for some $a \in \mathbb{Z}^m$ with $\gcd(a) = 1$, then there exists a *unimodular* matrix $Y = [a, X]^T$, and then $(YA)(1,1) = \alpha$. The same argument produces a unimodular matrix Z so that $(YAZ)(1,1) = \gcd(YA(1,:))$. Next, we obtain Y_1 so that $(Y_1YAZ)(1,1) = \gcd(YAZ(:,1))$, then Z_1 so that $(Y_1YAZZ_1)(1,1) = \gcd(Y_1YAZ(:,1))$, etc. At each step of this process, the element in the $(1,1)$ position is reduced, hence must eventually become stationary. At that point, it divides all elements in the first row and column, hence the addition of appropriate *integer* multiples of the first row (column) to all the other rows (columns) produces a matrix C with zeros in the first row and the first column, except for position $(1,1)$. Such elimination corresponds to multiplying from the left (right) by a unit triangular integer matrix, hence a unimodular matrix.

If now $C(1,1)$ fails to divide some $C(i,j)$, add column $C(:,j)$ to column 1 (which amounts to multiplication from the right by some unimodular matrix) and run the earlier process once again to completion. This will further decrease the element in position $(1,1)$. Since there are only finitely many entries in the matrix, repetition of this second process must therefore lead to a matrix $D = V^{-1}AU^{-1}$ (with U and V unimodular) for which $D(1,1) = \gcd(D)$ and whose first row and column is otherwise 0. Now induction provides the desired factorization for the submatrix $D(2{:}m, 2{:}n)$, hence ultimately for A. □

It would have been possible to use the ideas in the above proof to prove directly the following result which is the only one of real interest here. But this would not have saved much space or effort, and would have denied the reader the pleasure of seeing a proof of the rather remarkable Smith normal form.

(24)Corollary. *Any $A \in \mathbb{Z}^{m\times n}$ $(m \le n)$ of full rank can be written in the form $A = [X, 0]U$ with $X \in \mathbb{Z}^{m\times m}$ invertible and $U \in \mathbb{Z}^{n\times n}$ unimodular.*

Proof. The Smith normal form supplies unimodular matrices U and V so that $A = V\Sigma U$, with $\Sigma \in \mathbb{Z}^{m\times n}$ diagonal. If A is of full rank, then so must Σ be, hence $\Sigma = [D, 0]$ for some invertible diagonal integer matrix D. The conclusion now follows with $X := VD$. □

We now state, in the terms of the present discussion and for the reader's convenience, some notational conventions concerning submatrices described in Notations. For our given $A \in \mathbb{Z}^{m\times n}$, we denote by

$$A(P,Q)$$

the matrix made up of the rows indicated by P in $\{1,\ldots,m\}$ and the columns indicated by Q in $\{1,\ldots,n\}$. Further, $A(P,:) := A(P,\{1,\ldots,n\})$ and $A(:,Q) := A(\{1,\ldots,m\},Q)$. Also, when we write $\det A(P,Q)$, we assume that $\#P = \#Q$. For example, since our matrix A has m rows, the statement $\det A(:,Q)$ implies that Q is an m-sequence (with entries in $\{1,\ldots,n\}$, since our A has n columns).

When looking for integer solutions of the equation $A? = c$ for given $c \in \mathbb{Z}^n$, we are, in effect, asking whether c is in the subgroup or sublattice $A\mathbb{Z}^n$ generated by the integer matrix A. Not surprisingly, this will be the case if and only if the sublattice $[A,c]\mathbb{Z}^{n+1}$ of $A\mathbb{Z}^n$ is no finer than $A\mathbb{Z}^n$. One measure for the fineness of $A\mathbb{Z}^n$ is the number

$$d_A := \gcd\{\det A(:,Q) : Q \text{ in } \{1,\ldots,n\}\}$$

which gives the smallest volume of a parallelepiped spanned by points in the lattice $A\mathbb{Z}^n$. In particular, as we shall see, $A\mathbb{Z}^n = \mathbb{Z}^n$ if and only if $d_A = 1$. Since A is of full rank, d_A is a well-defined positive number. It is this number which plays the role of 'rank' of A when considering the diophantine equation $A? = c$. We make use of the following simple observation:

(25)Lemma. *If A, B, C are integer matrices with C square and $A = BC$, then $d_B \mid d_A$; hence, $d_B = d_A$ in case C is unimodular.*

Proof. That $d_B \mid d_A$, follows from the Cauchy-Binet formula

$$\det(BC(:,Q)) = \sum_P \det B(:,P) \det C(P,Q),$$

hence also $d_A \mid d_B$ in case C is unimodular, since then $B = AC^{-1}$ with C^{-1} integral. □

(26)Theorem. *The equation $A? = c$, with $A \in \mathbb{Z}^{m\times n}, c \in \mathbb{Z}^m$, has a solution in $\mathbb{Z}^n$ if and only if $d_A = d_{[A,c]}$, i.e., if and only if adjoining c to the columns of A does not refine the lattice generated by the matrix.*

Proof. If $Ax = c$, then $[A,c] = A[\mathbb{1},x]$. Therefore, by (25)Lemma, $d_A \mid d_{[A,c]}$ in case $x \in \mathbb{Z}^n$, while $d_{[A,c]} \mid d_A$ regardless.

For the converse, since A is of full rank, we obtain from (24)Corollary a unimodular matrix $U \in \mathbb{Z}^{n\times n}$ so that $A = [X,0]U$ for some invertible $X \in \mathbb{Z}^{m\times m}$. Hence $d_A = d_{[X,0]} = \det X$ by (25)Lemma. Since also

$$[A,c] = [X,0,c]\begin{bmatrix} U & 0 \\ 0 & 1 \end{bmatrix},$$

the same argument shows that $d_{[A,c]} = d_{[X,0,c]}$. Thus, if $d_A = d_{[A,c]}$, then also $\det X = d_{[X,0]} = d_{[X,0,c]}$, hence the vector $y := (X^{-1}c, 0)$ is in $\mathbb{Z}^n$ by Cramer's rule. Since U is unimodular, it follows that $x := U^{-1}y \in \mathbb{Z}^n$ and that $Ax = [X, 0]UU^{-1}(X^{-1}c, 0) = c$. ☐

(27)Corollary. *The integer matrix $A \in \mathbb{Z}^{m\times n}$ (with $m \le n$) maps $\mathbb{Z}^n$ onto $\mathbb{Z}^m$ if and only if $d_A = 1$.*

Proof. If $d_A = 1$, then $d_A \mid d_{[A,c]}$ for every $c \in \mathbb{Z}^m$. Conversely, if $A? = c$ has integer solutions for every $c \in \mathbb{Z}^m$, then, in particular, $AX = \mathbb{1}$ for some $X \in \mathbb{Z}^{n\times n}$, hence $d_A \mid d_{\mathbb{1}} = 1$, by (25)Lemma. ☐

With this, we are ready to discuss the linear independence question.

Linear independence (continued). Let

$$A := [\Xi^T, N].$$

We continue to denote by

$$A(P, Q) \tag{28}$$

the matrix made up of the rows indicated by P and the columns indicated by Q, in the order in which they appear in P and Q, and with the same multiplicity. The rows of our particular A are most conveniently indexed by $\xi \in \Xi$. Correspondingly, we use submatrices of Ξ for P in (28). The columns of A are of two kinds. The first s columns contain the entries of the $\xi \in \Xi$ and are most conveniently indexed by the integers $1, \ldots, s$. The remaining columns are most conveniently indexed by $\xi \in \Xi$. Thus Q in (28) is a subset of $\{1, \ldots, s\} \cup \Xi$. We use the abbreviations

$$[\mathrm{Z}^T, N(\mathrm{Z}, \mathrm{Z})] =: A(\mathrm{Z}), \qquad N_\mathrm{Z} := \prod_{\zeta\in\mathrm{Z}} N_\zeta.$$

In these terms, the following observation will be used repeatedly: If $\Theta \subseteq \mathrm{Z} \subseteq \Xi$ and $Q \subseteq \{1, \ldots, s\} \cup \Xi$, then

$$\det A(\Theta, Q \cup \mathrm{Z}) = \pm N_\mathrm{Z} \det A(\Theta\backslash\mathrm{Z}, Q). \tag{29}$$

We find the abbreviation

$$\textbf{1-1}(\Xi) := \{\mathrm{Z} \subseteq \Xi : \mathrm{Z}\ \text{1-1}\}$$

convenient for the statement and proof of the following characterization of the linear independence of the shifts of the discrete box spline b^h.

(30)Theorem. *Let $A = [\Xi^T, N]$. The following conditions are equivalent:*

(a) *The shifts of b^h are linearly independent.*

(b) *For all $\mathrm{Z} \in \mathbf{1\text{-}1}(\Xi)$,*

$$d_{A(\mathrm{Z},:)} = 1.$$

(c) *For all $\mathrm{Z} \in \mathbf{1\text{-}1}(\Xi)$,*

$$\gcd\{d_{\mathrm{Z}^T}, \gcd\{N_\zeta : \zeta \in \mathrm{Z}\}\} = 1.$$

Proof. Recall (from the discussion following (18)Result) that the linear independence of the shifts of b^h is equivalent to the existence of integer solutions $(j,k) \in \mathbb{Z}^s \times \mathbb{Z}^\Theta$ to the linear equations

$$A(\Theta)(j,k) = \Theta^T j + N(\Theta,\Theta)k = \Theta^T y \tag{31}$$

with $y \in \mathbb{C}^s$ arbitrary and $\Theta = \Theta_y$ consisting of those $\theta \in \Xi$ for which $\theta y \in \mathbb{Z}$. Since $A(\Theta)$ is of full rank, existence of an integer solution for (31) is, by (26)Theorem, equivalent to having $d_{A(\Theta)} \mid d_{[A(\Theta),\Theta^T y]}$, i.e., equivalent to having

$$d_{A(\Theta,:)} \mid d_{[A(\Theta,:),\Theta^T y]}, \tag{32}$$

since the columns of $N(\Theta,:)$ other than those in $N(\Theta,\Theta)$ are zero.

(a) $\Longrightarrow$ (b): Let $\mathrm{Z} \in \mathbf{1\text{-}1}(\Xi)$. Then Z^T is onto. In particular, we can write any $c \in \mathbb{Z}^{\mathrm{Z}}$ in the form $c = \mathrm{Z}^T y$ for some $y \in \mathbb{C}^s$. Therefore, by (26)Theorem, (a) implies the existence of integer solutions to $A(\mathrm{Z},:)? = c$ for arbitrary $c \in \mathbb{Z}^{\mathrm{Z}}$, and this implies $d_{A(\mathrm{Z},:)} = 1$, by (27)Corollary.

(b) $\Longrightarrow$ (c): Let $\mathrm{Z} \in \mathbf{1\text{-}1}(\Xi)$. We are to prove that

$$\gamma := \gcd\{d_{\mathrm{Z}^T}, \gcd\{N_\zeta : \zeta \in \mathrm{Z}\}\} = 1.$$

For this, note that any $Q \subseteq \{1,\ldots,s\} \cup \mathrm{Z}$ lies either entirely in $\{1,\ldots,s\}$ or else contains some $\zeta \in \mathrm{Z}$. In the first case, $\mathrm{Z}^T \supseteq A(\mathrm{Z},Q)$, hence $d_{\mathrm{Z}^T} \mid \det A(\mathrm{Z},Q)$. In the second case, we compute with (29) that

$$\det A(\mathrm{Z},Q) = \pm N_\zeta \det A(\mathrm{Z}\backslash\zeta, Q\backslash\zeta),$$

hence $N_\zeta \mid \det A(\mathrm{Z},Q)$. This implies that $\gamma \mid \gcd\{\det A(\mathrm{Z},Q) : Q\}$. Since

$$\gcd\{\det A(\mathrm{Z},Q) : Q\} = d_{A(\mathrm{Z},:)} = 1,$$

the second equality by (b), $\gamma = 1$ follows.

(c) $\Longrightarrow$ (b): We prove this by induction on #Z, it being vacuously true for #Z = 0. Since

$$\gcd\{d_{Z^T}, \gcd\{N_\zeta : \zeta \in Z\}\} = 1$$

by (c), and $d_{A(Z,:)} \mid d_{Z^T}$, it is sufficient to prove that $d_{A(Z,:)} \mid N_\zeta$ for all $\zeta \in Z$. But this follows from the fact that, for any $\zeta \in Z$,

$$\det A(Z, Q \cup \zeta) = \pm N_\zeta \det A(Z\backslash\zeta, Q)$$

by (29), hence

$$d_{A(Z,:)} \mid N_\zeta \gcd\{\det A(Z\backslash\zeta, Q) : \#Q = \#Z - 1\},$$

while

$$\gcd\{\det A(Z\backslash\zeta, Q) : \#Q = \#Z - 1\} = d_{A(Z\backslash\zeta,:)} = 1$$

by induction hypothesis.

(b) $\Longrightarrow$ (a): We are to prove (32), i.e.,

(33) $$d_{A(\Theta,:)} \mid d_{[A(\Theta,:),\Theta^T y]}$$

for arbitrary $y \in \mathbb{C}^s$ and with Θ consisting of those $\theta \in \Xi$ for which $\theta y \in \mathbb{Z}$. For this, recall that $d_{[A(\Theta,:),\Theta^T y]}$ is the gcd of all the numbers $\det B$, with B any square submatrix of order $\#\Theta$ of $[A(\Theta,:), \Theta^T y]$. It is sufficient to consider only submatrices B which include that last column (since $d_{A(\Theta,:)}$, by its definition, is the gcd of the determinants of all the others). Such a B is of the form $B = [A(\Theta, Q), N(\Theta, \Theta\backslash Z), \Theta^T y]$ for some Q in $\{1, \ldots, s\}$ and some $Z \subseteq \Theta$. Hence (with (29))

(34) $$\det B = \pm N_{\Theta\backslash Z} \det C,$$

with $C = [Z^T(:, Q), Z^T y] = Z^T[\mathbb{1}(:, Q), y]$. Consequently, if $Z \notin \mathbf{1\text{-}1}(\Theta)$, then $\det C = 0$, hence $\det B = 0$, and we can ignore such B. It follows that, for the B of interest, viz. those with $\det C \neq 0$, we must have $Z \in \mathbf{1\text{-}1}(\Theta)$, and we are done once we prove that, in that case,

(35) $$d_{A(\Theta,:)} \mid N_{\Theta\backslash Z},$$

since this implies with (34) that $d_{A(\Theta,:)} \mid \det B$. For the proof of (35), observe that any submatrix E of $A(Z,:)$ of order #Z can be extended to a submatrix D of $A(\Theta,:)$ of order $\#\Theta$ with $\det D = \pm N_{\Theta\backslash Z} \det E$. This shows that

$$d_{A(\Theta,:)} \mid d_{A(Z,:)} N_{\Theta\backslash Z},$$

and this proves (35) since $Z \in \mathbf{1\text{-}1}(\Theta) \subseteq \mathbf{1\text{-}1}(\Xi)$, and therefore $d_{A(Z,:)} = 1$, by (b). □

(36)Corollary. *If N is a scalar matrix, i.e., $h_\xi = 1/N$, all $\xi \in \Xi$, for some $N \in \mathbb{N}$, then the shifts of b^h are linearly independent if and only if N is relatively prime to every $\det \Theta$ with $\Theta \in \mathcal{B}(\Xi)$.*

Proof. In view of the preceding theorem, only the sufficiency needs proof. For this, observe that any $\mathrm{Z} \in \mathbf{1\text{-}1}(\Xi)$ can be extended to some $\Theta \in \mathcal{B}(\Xi)$, and, for any such extension, $d_{\mathrm{Z}^T} \mid \det \Theta$. Therefore, $\gcd\{d_{\mathrm{Z}^T}, N\} \mid \gcd\{\det \Theta, N\}$. This implies that condition (c) of the preceding theorem holds, since $\gcd\{\det \Theta, N\} = 1$, by assumption. ∎

On extending to an element of $\Delta(\Xi)$. Recall from (II.46)Theorem that any mesh function defined on $\iota(x, \Xi)$ for some $x \notin \Gamma(\Xi)$ has exactly one extension to an element of $\Delta(\Xi)$. For a more detailed look at the discrete box spline, we need to consider, more generally, the extension of a mesh function f given on

$$\iota(\Omega, \Xi) := (\Omega - \Xi\square) \cap \mathbb{Z}^s = \cup_{x \in \Omega}\, \iota(x, \Xi)$$

to an element of $\Delta(\Xi)$. If $\iota(\Omega, \Xi)$ is larger than one of the sets $\iota(x, \Xi)$, we cannot expect such an extension to work without restrictions on the given values on $\iota(\Omega, \Xi)$. These restrictions must ensure that $\Delta_{\mathrm{Z}} f = 0$ for all $\mathrm{Z} \in \mathcal{A}_{\min}(\Xi)$, to the extent that these difference operators can be applied to an f defined only on $\iota(\Omega, \Xi)$.

For this, it is convenient to deal, equivalently, with the backward difference operators ∇_{Z}, for the following reason. Since

$$\nabla_{\mathrm{Z}} \varphi = \sum_{Y \subseteq \mathrm{Z}} (-1)^{\# Y} \varphi(\cdot - \sum_{v \in Y} v), \tag{37}$$

computation of $\nabla_{\mathrm{Z}} f(j)$ requires the function values $f(j - \sum_{v \in Y} v)$ for all $Y \subseteq \mathrm{Z}$, i.e., requires values at the points $j - \mathrm{Z}\{0,1\}^{\mathrm{Z}}$. All these points are certain to lie in $\iota(\Omega, \Xi)$ in case $j \in \iota(\Omega, \Xi\backslash\mathrm{Z})$, i.e., $(j + (\Xi\backslash\mathrm{Z})\square) \cap \Omega \neq \emptyset$, since, for any $\alpha \in \mathrm{Z}\{0,1\}^{\mathrm{Z}}$,

$$j - \alpha + \Xi\square \supset j - \alpha + \alpha + (\Xi\backslash\mathrm{Z})\square = j + (\Xi\backslash\mathrm{Z})\square.$$

Thus we will assume that

$$\nabla_{\mathrm{Z}} f = 0 \quad \text{on } \iota(\Omega, \Xi\backslash\mathrm{Z}) \quad \forall \mathrm{Z} \in \mathcal{A}_{\min}(\Xi).$$

It is not clear *a priori* that this covers *all* conditions of the form

$$\Delta_{\mathrm{Z}} f(j) = 0$$

for a $\mathrm{Z} \in \mathcal{A}_{\min}(\Xi)$ for which $j + \mathrm{Z}\{0,1\}^{\mathrm{Z}} \subset \iota(\Omega, \Xi)$.

If Ω is an arbitrary set, then these conditions may not be sufficient to ensure extendibility. The following Proposition asserts that they are sufficient in case Ω is connected.

(38)Proposition. *For a nonempty connected* Ω,
(39)
$$\Delta(\Xi)_{|\iota(\Omega,\Xi)} = \{f : \iota(\Omega,\Xi) \to \mathbb{C} : \nabla_{\Xi\backslash H} f(x) = 0 \text{ on } \iota(\Omega, \Xi\cap H), H\in\mathbb{H}(\Xi)\}.$$

Proof. We begin with the observation that we may assume without loss of generality that Ω is not only connected, but also open. For, with ω an arbitrary point in Ω, $\omega-\Xi\square$ is closed and bounded, hence the distance of $\omega-\Xi\square$ from $\mathbb{Z}^s\backslash(\Omega-\Xi\square)$ is positive, and so $\iota(\omega,\Xi) = \iota(B_r(\omega),\Xi)$ for some positive $r = r(\omega)$. Therefore, $\iota(\Omega,\Xi) = \iota(\Omega',\Xi)$, with $\Omega' := \cup_{\omega\in\Omega} B_{r(\omega)}(\omega)$ an open connected set.

Let F denote the right-hand side of (39). We only need to prove that any $f \in F$ is the restriction to $\iota(\Omega,\Xi)$ of some $g \in \Delta(\Xi)$. We know from (II.46)Theorem that, for every $x \in \Omega\backslash\Gamma(\Xi)$, there is exactly one $g_x \in \Delta(\Xi)$ which agrees with our f on $\iota(x,\Xi)$. Thus our task is to show that all these g_x coincide. Since $\iota(x,\Xi)$ is the same set for each x in any given connected component c of $\backslash\Gamma(\Xi)$, the corresponding extension g_x is the same function for every $x \in c$. Call this function g_c. Thus we must show that $g_c = g_{c'}$ for any two connected components c, c' of $\backslash\Gamma(\Xi)$ which meet Ω. Since Ω is open and connected, there exists a finite sequence $\omega_1, \omega_2, \ldots, \omega_\ell$ with $\omega_1 \in c$, $\omega_\ell \in c'$, and so that, for each j, $[\omega_j \,.\,.\, \omega_{j+1}]$ is not only in Ω, but is either entirely contained in a connected component of $\backslash\Gamma(\Xi)$ or else is contained (except for one point) in the union of two connected components separated by a hyperplane. This implies that it is sufficient to prove that $g_c = g_{c'}$ for any c, c' separated by a hyperplane, e.g., by the hyperplane $v + H$ for some $v \in \mathbb{Z}^s$ and some $H \in \mathbb{H}(\Xi)$.

Now consider $\iota(x,\Xi)$ as x moves from c to c', along a straight line. At the moment of crossing of the hyperplane $v + H$, at x_t say, all the points of
$$\iota(c \cup c', \Xi) = \iota(c,\Xi) \cup \iota(c',\Xi)$$
lie in the (closed) support of $M_\Xi(x_t - \cdot)$, while those points in one but not the other will leave that support as soon as we move away from x_t into the interior of c or c'. This implies that all the points in $I := \iota(c,\Xi)\backslash\iota(c',\Xi)$ lie on some shift of the hyperplane H, while all the points in $I' := \iota(c',\Xi)\backslash\iota(c,\Xi)$ lie on some other shift of H, and that both I and I' lie in the boundary of the support of $M_\Xi(x_t - \cdot)$, i.e., of $M_\Xi(\cdot - y_t)$, with $y_t = x_t - 2c_\Xi$ (the last statement uses (I.20) and (I.23)).

Now recall from Chapter I, esp. (I.54), (I.56), that the boundary points of the support of M_Ξ associated with the hyperplane $H \in \mathbb{H}(\Xi)$ must lie in the set
$$H + \sum_{\zeta\in \mathrm{Z}} a(\zeta)\zeta,$$
with
$$a(\zeta) := \begin{cases} 1, & \text{if } H^\perp\zeta/H^\perp\eta > 0; \\ 0, & \text{if } H^\perp\zeta/H^\perp\eta < 0, \end{cases} \qquad \text{for } \zeta \in \mathrm{Z} := \Xi\backslash H,$$

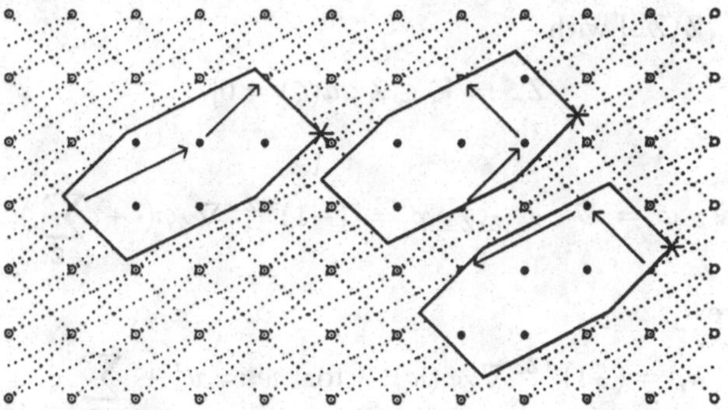

(40)Figure. Three examples of $\iota(x,\Xi)$ for $x \in \Gamma$, showing how corresponding mesh points on the two boundaries parallel to the hyperplane H differ by $\sum_{\xi\in\Xi\backslash H} \pm\xi$. Specifically, $\Xi = \left[\begin{smallmatrix} 1 & 2 & 1 \\ -1 & 1 & 1 \end{smallmatrix}\right] =: [\zeta, \eta, \vartheta]$, and, from left to right, $H = \operatorname{ran}[\zeta]$, $\operatorname{ran}[\eta]$, $\operatorname{ran}[\vartheta]$.

where $H^\perp$ is a (nontrivial) normal to H and η is any vector pointing from the boundary point in question to the outside of the support of M_Ξ. In particular, with $w' \in I'$, let a be the vector so associated with the boundary point $w' - y_t$ of the support of M_Ξ. Set

$$\mathrm{Z}_\varepsilon := \mathrm{Z}\operatorname{diag}\varepsilon, \quad \text{with } \varepsilon := 2a - 1 \in \{-1, 1\}^s.$$

Then, by (37), the linear functional

$$\lambda : \varphi \mapsto \nabla_{\mathrm{Z}_\varepsilon} \varphi(w')$$

is of the form

$$\lambda\varphi = \sum_{v\in V} \varphi(v)\mu_v,$$

with

$$V := w' - \{\sum_{v\in Y} v : Y \subseteq \mathrm{Z}_\varepsilon\}$$

in the closed support of $M_\Xi(\cdot - y_t)$, hence in $\iota(c \cup c', \Xi)$. Further, w' is the only point in V not in $\iota(c, \Xi)$, and has the weight $\mu_{w'} = 1$ in λ. Since $f = g_c$ on $\iota(c, \Xi)$, this implies that

$$(41) \qquad (f - g_c)(w') = \lambda(f - g_c).$$

Now, since

$$\nabla_{-z}\varphi = \varphi - \varphi(\cdot + z) = -(\varphi(\cdot + z) - \varphi(\cdot + z - z)) = -\nabla_z\varphi(\cdot + z),$$

and $Z = [Z_-, Z\backslash Z_-]$ with

$$Z_- := [\zeta \in Z : a(\zeta) = 0],$$

we have

$$\nabla_{Z_\epsilon}\varphi = \nabla_{[-Z_-, Z\backslash Z_-]}\varphi = (-1)^{\#Z_-}\nabla_Z\varphi(\cdot + \sum_{\zeta\in Z_-}),$$

and therefore

$$\lambda\varphi = (-1)^{\#Z_-}\nabla_Z\varphi(w), \quad \text{for } w := w' + \sum_{\zeta\in Z_-}\zeta.$$

Thus, if $w \in \iota(x_t, \Xi \cap H)$, then we know that $\lambda f = 0$, while $\nabla_Z g_c(j) = 0$ for any $j \in \mathbb{Z}^s$, hence in particular $\lambda g_c = 0$. Consequently, $\lambda(f - g_c) = 0$, and this implies with (41) that $f = g_c$ at w'. Since $w' \in I'$ was arbitrary here, we have $f = g_c$ not only on $\iota(c, \Xi)$, but also on any point $w' \in \iota(c', \Xi)\backslash\iota(c, \Xi)$. Since $g_{c'} = f$ on $\iota(c', \Xi)$, it follows that $g_c = g_{c'}$ on $\iota(c', \Xi)$, hence $g_c = g_{c'}$, by (II.46)Theorem.

It remains to verify that $w \in \iota(x_t, \Xi \cap H)$. For this, observe that

$$w' - y_t \in (\Xi \cap H)\square + \sum_{\zeta\in Z\backslash Z_-}\zeta,$$

hence

$$w - x_t = w' + \sum_{\zeta\in Z_-}\zeta - (y_t + \sum_{\xi\in\Xi}\xi) \in (\Xi\cap H)\square + \sum_{\xi\in\Xi\cap H}\xi,$$

which shows that $x_t - w$ lies in the closed support of $M_{\Xi\cap H}$. □

Discrete truncated power. Assume that 0 is an extreme point of $\Xi\square$ (which can always be achieved by an appropriate shift; see the discussion leading up to (I.25)). Then $M_\Xi = \nabla_\Xi T_\Xi$, with T_Ξ the corresponding truncated power, i.e., (recalling from (I.26)) the distribution given by

$$\langle T_\Xi, \varphi\rangle = \int_{\mathbb{R}^n_+}\varphi(\Xi t)dt, \qquad \varphi \in C_0(\mathbb{R}^s).$$

On the other hand,

$$\begin{aligned}\langle T_\Xi, \varphi\rangle &= \int_{\mathbb{R}^n_+}\varphi(\Xi t)dt \\ &= \sum_{\alpha\in\mathbb{Z}^n_+}\int_{\alpha+\llcorner}\varphi(\Xi t)dt \\ &= \sum_{\alpha\in\mathbb{Z}^n_+}\langle M_\Xi(\cdot - \Xi\alpha), \varphi\rangle,\end{aligned}$$

hence

$$T_\Xi = M_\Xi * t_\Xi,$$

with

$$t_\Xi(j) := \#\{\alpha \in \mathbb{Z}_+^n : \Xi\alpha = j\}, \qquad j \in \mathbb{Z}^s$$

the **discrete truncated power** determined by Ξ.

The support of T_Ξ consists of the cone $\Xi\mathbb{R}_+^n$, and this cone is the essentially disjoint union of the closures of the so-called **fundamental Ξ-cones.** These are the connected components of $\Xi\mathbb{R}_+^n\backslash\Gamma_0(\Xi)$, with

$$\Gamma_0(\Xi) := \bigcup_{H\in\mathbb{H}(\Xi)} H$$

the union of the 'mesh-planes' for Ξ through the origin. On each fundamental Ξ-cone, T_Ξ coincides with some homogeneous element of $D(\Xi)$ of maximal degree, i.e., of degree $k(\Xi)$, since T_Ξ is homogeneous of that degree. We claim that the discrete truncated power is similarly structured, being piecewise in $\Delta(\Xi)$.

(42)Theorem. *For any fundamental Ξ-cone Ω, t_Ξ coincides on $\Omega - \Xi\square$ with some $f \in \Delta(\Xi)$. For any $x \in \Omega$ with $\iota(x,\Xi) \cap \mathbb{Z}_+^n = \{0\}$, f is the unique element of $\Delta(\Xi)$ which agrees with δ on $\iota(x,\Xi)$.*

Proof. Let $Z \in \mathcal{A}(\Xi)$ and $j \in \iota(\Omega - \Xi\square, \Xi\backslash Z)$. If $j \in \operatorname{ran}(\Xi\backslash Z)$, then the fact that $(j + 2(\Xi\backslash Z)\square) \cap \Omega \neq \emptyset$ would imply that $\operatorname{ran}(\Xi\backslash Z) \cap \Omega \neq \emptyset$, thus contradicting the assumption that Ω is a fundamental Ξ-cone. Consequently, any such j is not in $\operatorname{ran}(\Xi\backslash Z)$, and therefore $t_{\Xi\backslash Z}(j) = 0$. This proves that $\nabla_Z t_\Xi = 0$ on $\iota(\Omega - \Xi\square, \Xi\backslash Z)$ for any $Z \in \mathcal{A}(\Xi)$, hence (38)Proposition implies the existence of a unique $f \in \Delta(\Xi)$ which agrees with t_Ξ on $\Omega - \Xi\square$. Further, for any $x \in \Omega$ with $\iota(x,\Xi) \cap \Xi\mathbb{R}_+^n = \{0\}$, we have $t_\Xi = \delta$ on $\iota(x,\Xi)$, hence also $f = \delta$ on $\iota(x,\Xi)$. ☐

The local structure of the discrete box spline. The local structure of the discrete box spline $b^h = b_\Xi^h$ reflects in the same way the local structure of the box spline $M = M_\Xi$.

To recall from (I.37)Proposition, let $\Gamma_{\text{loc}}(\Xi)$ be the mesh for $M = M_\Xi$, i.e.,

$$\Gamma_{\text{loc}}(\Xi) := \{\Xi t : \ \{\xi \in \Xi : t_\xi \in \{0,1\}\} \in \mathcal{A}(\Xi)\ \}.$$

Then, on each connected component of $(\Xi\square)\backslash\Gamma_{\text{loc}}(\Xi)$, M agrees with some element of $D(\Xi)$. Here is the corresponding result for the discrete box spline. It concerns functions f defined on any lattice $\mathbb{M}$ which contains the points $\Xi_h\mathbb{Z}^n$ and, in particular, such functions in

$$\Delta(\Xi_h) := \bigcap_{Z\in\mathcal{A}(\Xi)} \ker \nabla_{Z_h} = \{f : \mathbb{M} \to \mathbb{C} : \nabla_{Z_h} f = 0 \ \ \forall Z \in \mathcal{A}(\Xi)\}.$$

(43) Theorem. *For each connected component* Ω *of* $(\Xi\square)\backslash\Gamma_{\rm loc}(\Xi)$, *there exists a unique* $f \in \Delta(\Xi_h)$ *which agrees with* b^h_Ξ *on* $\Omega - \Xi_h\square$.

Proof. From (17),

$$b^h_\Xi = b^h_Z * b^h_{\Xi\backslash Z}$$

for any $Z \subseteq \Xi$, hence

$$\nabla_{Z_h} b^h_\Xi = h_Z \nabla_Z b^h_{\Xi\backslash Z}, \tag{44}$$

with $h_Z := \prod_{\zeta\in Z} h_\zeta$, by an argument which parallels that for (13). If now $Z \in \mathcal{A}(\Xi)$ and $y \in \Omega - \Xi_h\square$, then $y \notin \mathrm{ran}(\Xi\backslash Z)$, for, otherwise $\Omega \cap \mathrm{ran}(\Xi\backslash Z) \neq \emptyset$, a contradiction. Consequently, $\nabla_{Z_h} b^h_\Xi = 0$ on $\Omega - \Xi_h\square$ for any $Z \in \mathcal{A}(\Xi)$, and this finishes the proof, by (38)Proposition. ∎

Local linear independence. We say that the shifts of b^h are **locally** linearly independent if, for any open set Ω, the shifts $b^h(\cdot - j)$, $j \in \iota(\Omega, \Xi)$, are linearly independent on

$$\iota_h(\Omega, \Xi) := \{x \in \mathbb{R}^s : (x + \Xi_h\square) \cap \Omega \neq \emptyset\}.$$

(45) Theorem. *The shifts of* b^h *are locally linearly independent if and only if they are linearly independent.*

Proof. For given $\Omega \subseteq \mathbb{R}^s$, consider the set of indices j for which $b^h(\cdot - j)$ has some support in $\iota(\Omega, \Xi_h)$, i.e., the set

$$\iota'_h(\Omega, \Xi) := \{j \in \mathbb{Z}^s : (j + \mathrm{supp}\, b^h) \cap \iota(\Omega, \Xi_h) \neq \emptyset\}.$$

We claim that

$$\iota'_h(\Omega, \Xi) \subseteq \iota(\Omega, \Xi). \tag{46}$$

(In fact, equality holds; but we don't need that.) For this, observe that, by (16) and the nonnegativity of M and b^h,

$$\mathrm{supp}\, b^h \subseteq \{k \in \Xi_h\mathbb{Z}^n : k + \Xi_h\square \subset \Xi\square\}.$$

Thus, if $j \in \iota'_h(\Omega, \Xi)$, then $k - j + \Xi_h\square \subseteq \Xi\square$ for some $k \in \iota(\Omega, \Xi_h)$, i.e.,

$$k + \Xi_h\square \subseteq j + \Xi\square.$$

Since $k \in \iota(\Omega, \Xi_h)$, $(k + \Xi_h\square) \cap \Omega \neq \emptyset$, hence also $(j + \Xi\square) \cap \Omega \neq \emptyset$, i.e., $j \in \iota(\Omega, \Xi)$.

To prove local linear independence of the shifts of b^h, it will suffice to prove that, for any $x \in \mathbb{R}^s \backslash \Gamma(\Xi)$, the shifts $b^h(\cdot - j)$ with $j \in \iota'_h(x, \Xi)$ are linearly independent on $\iota(x, \Xi_h)$. Assume that the shifts of b^h are linearly independent. Then the linear map

$$b^h *' : f \mapsto \sum_{j \in \mathbb{Z}^s} b^h(\cdot - j) f(j)$$

on $\mathbb{R}^{\mathbb{R}^s}$ is 1-1. Further, by (44),

$$\nabla_{Z_h}(b^h *' f) = h_Z(\nabla_Z b^h_{\Xi \backslash Z}) *' f = h_Z b^h_{\Xi \backslash Z} *' \nabla_Z f,$$

hence $b^h *'$ maps $\Delta(\Xi)$ into $\Delta(\Xi_h)$ and does so in a 1-1 manner. Consequently,

$$\dim \Delta(\Xi) = \dim(b^h *'(\Delta(\Xi))) = \dim \left(b^h *'(\Delta(\Xi))\right)_{|\iota(x, \Xi_h)},$$

the last equality by (38)Proposition, while, by (46) and (II.46)Theorem,

$$\# \iota'_h(x, \Xi) \le \# \iota(x, \Xi) = \dim \Delta(\Xi),$$

and, by the definition of $\iota'_h(x, \Xi)$,

$$\dim \left(b^h *'(\Delta(\Xi))\right)_{|\iota(x, \Xi_h)} \le \# \iota'_h(x, \Xi) = \#\{j : \operatorname{supp} b^h(\cdot - j) \cap \iota(x, \Xi_h) \ne \emptyset\}.$$

This implies that $b^h(\cdot - j)$, $j \in \iota'_h(x, \Xi)$, are linearly independent on $\iota(x, \Xi_h)$. ▯

Notes. The *discrete box spline* was introduced by Cohen, Lyche, and Riesenfeld in [Cohen, Lyche, Riesenfeld'84], and by Dahmen and Micchelli [Dahmen, Micchelli'84a], who were following up on ideas in [Prautzsch'84c] concerning subdivision for three-direction box splines.

There are (at least) two useful normalizations of the discrete box spline. The one we have chosen in this chapter (and denoted by b^h_Ξ) parallels the (standard) normalization of the (continuous) box spline in that b^h_Ξ converges to M_Ξ in the sense of distributions, as $h \to 0$. For computational use (as detailed in Chapter VII), it is more advantageous to think of the discrete box spline as a mesh function or mask, with the connection to functions defined on all of $\mathbb{R}^s$ made by restriction of the latter to the appropriate mesh. If we want the discrete box spline to relate properly in this way to M_Ξ, then the normalization $m^h := b^h / h^s$ is appropriate, and we use this throughout Chapter VII.

The description given here of the basic properties of the discrete box spline has benefitted from suggestions made by Kang Zhao, especially as concerns (11), (13), and (14).

Starting with the section on *linear independence*, the rest of the chapter is based on various versions of [Jia'9x] (in particular, but not exclusively, on [Jia'91a]) which, in turn, is a follow-up on papers of Dahmen and Micchelli.

In particular, the connection of box splines to *linear diophantine equations* was made by Dahmen and Micchelli, in [Dahmen, Micchelli'88c], where they studied the discrete truncated power as a tool to reprove and extend some results of Stanley concerning magic squares. Using these techniques, [Jia'91a,'91b] established some conjectures of Stanley (made in [Stanley'73]; see, e.g., [Stanley'86]) concerning the number of symmetric magic squares. For it, Jia needed extensions of the Dahmen-Micchelli results on the structure of the discrete truncated power, and it is these extensions that are reported in the present chapter.

The question of *linear independence* of the shifts of a discrete box spline was first raised in [Dahmen, Micchelli'87a].

The material on the Smith normal form is adapted from [Newman'72]. (26)Theorem and its corollary are due to [Skolem'50].

The characterization of *linear independence* of discrete box spline shifts, (30)Theorem, comes from [Jia'91a,'9x]. To compare, [Dahmen, Micchelli '87a] showed, for the special case of *uniform scaling*, that unimodularity of Ξ is sufficient but not necessary for the linear independence. They proved the sufficiency part of (36)Corollary under an additional condition.

(38)Proposition is Jia's extension (in [Jia'91a,'9x]) of Dahmen and Micchelli's result recorded in (II.46)Theorem. Jia's proof of it makes no use of (II.46)Theorem nor of facts about the support of a box spline, but uses instead new results concerning the solution of systems of difference equations. Since (II.46)Theorem is a special case of (38)Proposition, and has the fact that $\operatorname{vol}\Xi\square = \dim\Delta(\Xi)$ as an immediate consequence, this provides an alternative avenue to the calculation of $\operatorname{vol}\Xi\square$.

The *discrete truncated power* was introduced in [Dahmen, Micchelli'84f] and was shown there to provide the correct coefficients in the box spline series for Dahmen's truncated power. It was used by Dahmen and Micchelli, in [Dahmen, Micchelli'86d,'87a], in their study of *the structure of the discrete box splines*. In particular, [Dahmen, Micchelli'87a] contains (43)Theorem for the special case of a uniform scaling for which, moreover, $1/h$ is relatively prime to every $\det \mathrm{B}$ with $\mathrm{B} \in \mathcal{B}(\Xi)$. The full (43)Theorem is taken from [Jia'91a,'9x], as is (45)Theorem, on the *local linear independence* of the shifts of an arbitrary discrete box spline.

Finally, [Jia'9x] contains extensions of these results to the exponential box splines of Ron (cf. [Ron'88], [Dahmen, Micchelli'89a]).

VII

Subdivision algorithms

Since the shifts of the box spline $M = M_\Xi$ form a nonnegative, local partition of unity, $(M*a)(x)$ is a finite convex combination of the coefficients a, i.e.,

(1) $$(M*a)(x) = \sum_{j\in\mathbb{Z}^s} M(x-j)a(j) \in \operatorname{conv}\{a(j) : j \in \iota(x,\Xi)\}.$$

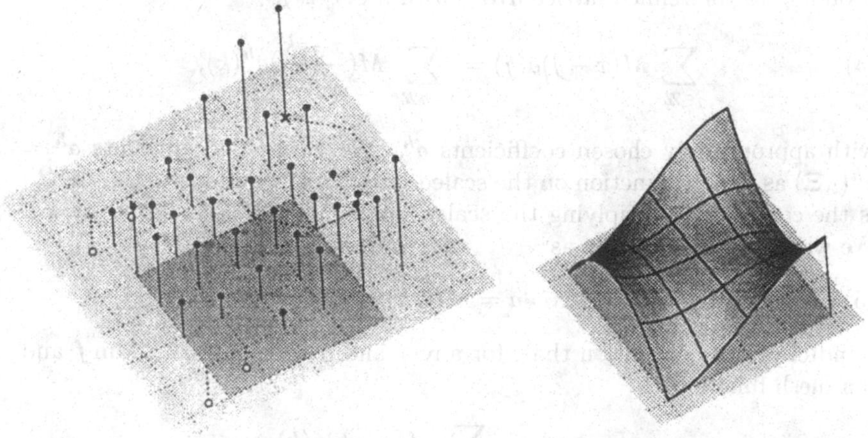

(2)Figure. Part of a bivariate cardinal spline (based on the ZP element) and the relevant box spline coefficients, i.e., those coefficients for which the support of the corresponding shifted box spline overlaps the domain on which the spline is plotted.

In particular, to the extent that the local variation of the coefficients is small, the mesh function a approximates well to the cardinal spline $M{*}a$. This is the basis for fast algorithms for graphic display of cardinal splines and rendering of box spline surfaces which are described in this chapter. For the discussion of subdivision algorithms, we assume throughout that $\operatorname{ran}\Xi = \mathbb{R}^s$.

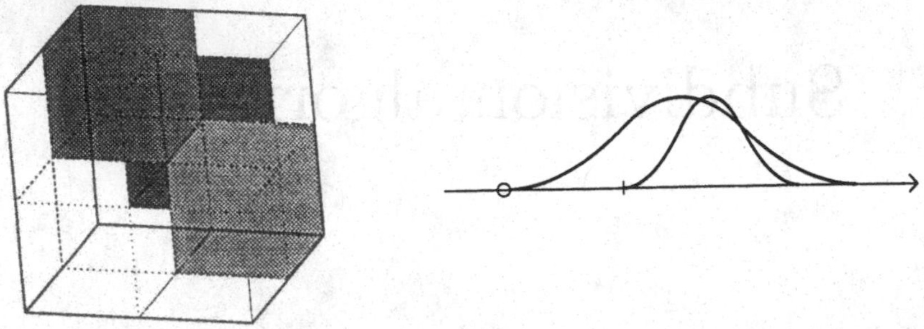

(3)Figure. Subdivision, with $h = 1/2$, of a (univariate) box spline. The figure highlights the three boxes which correspond to the same scaled box spline $M(\cdot/h - 2)$, also shown, in the representation of M.

Masks. The key observation is that the variation of the coefficients is reduced when a cardinal spline is expressed in terms of box splines corresponding to the refined lattice $h\mathbb{Z}^s$ (with $h \in 1/\mathbb{N}$),

$$\sum_{j\in\mathbb{Z}^s} M(x-j)a(j) = \sum_{k\in h\mathbb{Z}^s} M(\frac{x-k}{h})a^h(k), \tag{4}$$

with appropriately chosen coefficients a^h. We view the coefficients $a^h = a^h(\cdot,\Xi)$ as a mesh function on the scaled lattice $h\mathbb{Z}^s$ to indicate that $a^h(k)$ is the coefficient multiplying the scaled box spline $M(\cdot/h)$ shifted by kh. We write (4) more simply as

$$M{*}a = M(\cdot/h){*}a^h, \tag{5}$$

mindful of our convention that, for any f and g with $\operatorname{dom} g \subseteq \operatorname{dom} f$ and g a mesh function,

$$f{*}g : x \mapsto \sum_{k\in\operatorname{dom} g} f(x-k)g(k).$$

Since box spline shifts are in general not linearly independent, (5) does not determine a^h uniquely. But (VI.10) suggests the choice

$$a^h := m^h{*}a, \tag{6}$$

with the **mask** m^h the properly scaled discrete box spline,

$$m^h := b^h/h^s. \tag{7}$$

For, according to (VI.10), we have the **refinement equation**

$$M = M(\cdot/h)*m^h, \tag{8}$$

and therefore

$$M*a = (M(\cdot/h)*m^h)*a = M(\cdot/h)*(m^h*a). \tag{9}$$

By our convention, the range of summation for the above convolutions is the common domain of definition for the operands. Therefore, written out in detail, the identity (9) reads

$$\begin{aligned}\sum_{j\in\mathbb{Z}^s} M(\cdot - j)a(j) &= \sum_j \sum_{k\in h\mathbb{Z}^s} M(\frac{\cdot - j - k}{h})m^h(k)a(j)\\ &= \sum_k M(\frac{\cdot - k}{h})\sum_j m^h(k-j)a(j).\end{aligned}$$

Geometric derivation of the refinement equation. It may be instructive to rederive (8) here entirely in geometric terms as illustrated in (3)Figure. We partition the box $\square$ into h^{-n} boxes of side length h,

$$\square = \bigcup_{\alpha\in\square_h} \alpha + h\square,$$

with

$$\square_h = \{0, h, 2h, \ldots, 1-h\}^n = \square\cap h\mathbb{Z}^n.$$

(As with $\square$, we will consider the dimension of $\square_h$ to be determined from the context. For example, $Z\square_h = Z\{0, h, 2h, \ldots, 1-h\}^{\#Z}$.)

With this definition, (I.3) yields

$$\begin{aligned}M(x) &= \sum_{\alpha\in\square_h} \mathrm{vol}_{n-s}\{\Xi^{-1}x\cap(\alpha+h\square)\}/|\det\Xi|\\ &= \sum_{\alpha\in\square_h} \mathrm{vol}_{n-s}\{(\Xi^{-1}x-\alpha)\cap h\square\}/|\det\Xi|.\end{aligned}$$

Since $\Xi^{-1}x - \alpha = \Xi^{-1}(x-k)$ in case $\Xi\alpha = k$, we can combine all terms whose α is related in this way to the same $k \in h\mathbb{Z}^s$. Therefore, since $\Xi\alpha \in h\mathbb{Z}^s$ for any $\alpha \in \square_h$,

$$M(x) = \sum_{k\in h\mathbb{Z}^s} \#\{\alpha\in\square_h : \Xi\alpha = k\}\, \mathrm{vol}_{n-s}\{(\Xi^{-1}(x-k)\cap h\square\}/|\det\Xi|.$$

On the other hand, also from (I.3),

$$M((x-k)/h) = \mathrm{vol}_{n-s}\{\Xi^{-1}((x-k)/h) \cap \square\}/|\det \Xi|$$
$$= h^{s-n}\, \mathrm{vol}_{n-s}\{\Xi^{-1}(x-k) \cap h\square\}/|\det \Xi|,$$

using the fact that a scaling by h scales the $(n-s)$-dimensional volume by h^{n-s}. This proves that the identity (8) holds with

$$(10) \qquad m^h(k) = h^{n-s}\#\{\alpha \in \square_h : \Xi\alpha = k\},$$

i.e., (cf. (VI.3)) with $m^h = b^h/h^s$.

Factorization of the mask. We rewrite (10) in the following suggestive form

$$(11) \qquad m^h(k) = h^n \sum_{\alpha\in\square_h} \delta(k - \Xi\alpha)/h^s.$$

See (VI.1) and (VI.8) for specific examples of such a discrete box spline. It follows from (11) that the coefficients a^h defined in (6) satisfy

$$(12) \qquad a^h(k) = \sum_{j\in\mathbb{Z}^s} m^h(k-j)a(j) = h^{n-s} \sum_{\alpha\in\square_h} a(k - \Xi\alpha).$$

In geometric terms, $a^h(k)$ is a weighted sum of those coefficients $a(j)$ for which $j \in \mathbb{Z}^s$ also lies on the refined lattice $k - \Xi\square_h$, as is illustrated in (13)Figure. The relevant lattice points are shown as black dots in (13)Figure. The weights are the (suitably scaled) values of the discrete box spline.

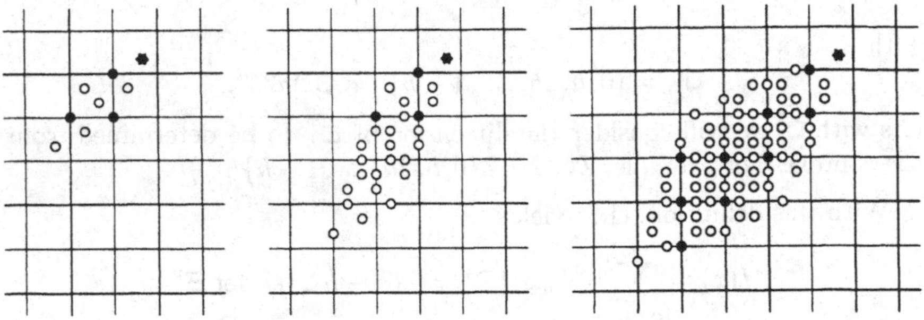

(13)Figure. The intersection of the set $k - \Xi\square_h$ with the integer lattice $\mathbb{Z}^s$ for $h = 1/3$ and $\Xi = \begin{bmatrix} 2 & 1 \\ 1 & 2 \end{bmatrix}, \begin{bmatrix} 2 & 1 & 1 \\ 1 & 2 & 3 \end{bmatrix}, \begin{bmatrix} 1 & 2 & 1 & 3 \\ 2 & 1 & 3 & 1 \end{bmatrix}$.

In view of (6) and (7), the inductive construction of b^h via (VI.5) yields a corresponding algorithm for computing a^h from $a =: a^1$. More generally, we conclude from (VI.11) that

$$(14) \qquad a^{hh'} = (\sigma_{h'}m^h)*a^{h'} = \sigma_{h'}(m^h*\sigma_{1/h'}a^{h'}),$$

(with $\sigma_h \varphi = \varphi(\cdot/h)$). The repeated application of this identity, i.e., the successive computation of a sequence of refined coefficients $a, a^{h'}, a^{hh'}, \ldots$, is called a **subdivision algorithm** in order to emphasize the geometric interpretation of the identities involved. A particularly simple form of the subdivision process results if one also chooses $h = h' = \cdots$, since then the subdivision process amounts to a repetition of the same simple averaging step(s). The **subdivision mask** m^h has smallest support when $h = 1/2$. A few examples are shown in (VI.8)Figure.

The convolution in (14) can be carried out by successive averaging since, in view of (VI.5),

(15) $$m^h_\Xi = b^h_\xi * b^h_\eta * \cdots * \delta/h^s,$$

if $\xi, \eta, \ldots$ are the directions in Ξ, and, e.g.,

$$b^h_\xi * c = h \sum_{j=0}^{1/h-1} c(\cdot - (jh)\xi)$$

(see (VI.7)Figure for an illustration). This leads to the following implementation of the subdivision step $a^{h'} \mapsto a^{hh'}$.

(16)Algorithm. *To compute $a^{hh'}$ from $a^{h'}$, set*

(i) $c(k) := a^{h'}(h'k)/h^s, \quad k \in h\mathbb{Z}^s,$

(keeping in mind our convention that a mesh function takes the value 0 on any point not in its domain). *Then, average the mesh function c in each of the directions in Ξ, i.e., for each $\xi \in \Xi$ compute*

(ii) $c \leftarrow b^h * c = h\Big(c + c(\cdot - h\xi) + \ldots + c(\cdot - (1-h)\xi)\Big)$

and, after that, define $a^{hh'}(k) := c(k/h')$ for $k \in h'h\mathbb{Z}^s$.

This algorithm is illustrated in the (17)Figure below. In step (i), $a^{h'}(h'k) = 0$ if $k \notin \mathbb{Z}^s$. Hence, in defining c on the lattice $h\mathbb{Z}^s$, the values of $a^{h'}$ (multiplied by h^{-s}) are assigned to c on the sublattice $\mathbb{Z}^s \subset h\mathbb{Z}^s$, as is indicated in the figure. In effect, c is the result of convolving $a^{h'}(\cdot h')$ with the Dirac sequence (multiplied by h^{-s}) on the fine mesh $h\mathbb{Z}^s$, in accordance with the last convolution in (15). The figure also shows lattice points involved in step (ii). In the final step, the mesh function c is transformed to the fine lattice $h'h\mathbb{Z}^s$ by scaling.

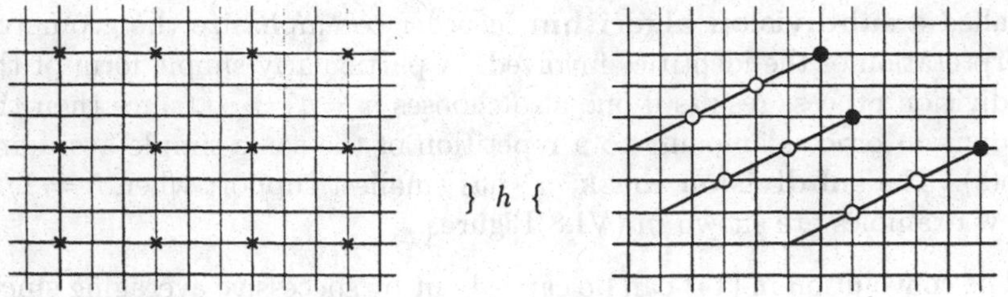

(17)Figure. Steps of the subdivision algorithm.

Again, (16) becomes particularly simple if $h = 1/2$. In this case, the averaging step becomes

(ii)′ $c(k) \leftarrow \bigl(c(k) + c(k - \xi/2)\bigr)/2$.

Example: two-direction mesh. Let $\Xi = \left[\begin{smallmatrix} 2 & -1 \\ 1 & 1 \end{smallmatrix}\right]$ and $h = 1/2$. To illustrate steps (i) and (ii)′ of (16)Algorithm, we denote adjacent values of the input mesh function by the letters $z, y, x, \ldots$. Then, step (i) can be symbolically written as

$$\cdots \begin{array}{ccc} z & y & x \\ w & v & u \end{array} \cdots \longrightarrow \cdots \begin{array}{ccccc} 4z & 0 & 4y & 0 & 4x \\ 0 & 0 & 0 & 0 & 0 \\ 4w & 0 & 4v & 0 & 4u \end{array} \cdots,$$

and step (ii)′ as

$$\cdots \begin{array}{ccccc} 4z & 0 & 4y & 0 & 4x \\ 0 & 0 & 0 & 0 & 0 \\ 4w & 0 & 4v & 0 & 4u \end{array} \cdots \xrightarrow[(2,1)]{} \cdots \begin{array}{ccc} 2y & 0 & 2x \\ 2w & 0 & 2v \end{array} \cdots \xrightarrow[(-1,1)]{} \cdots \; y \quad v \; \cdots .$$

Subdivision as discrete smoothing. The subdivision process is a discrete smoothing procedure, i.e., it reduces the variation of the coefficient sequence. This is made more precise in the following theorem, in which $\|a\|$ denotes the max-norm of the mesh function a.

(18)Theorem. *If* $\Xi \backslash Z$ *spans, then*

$$\|\nabla_{hZ} a^h\| \le \text{const}_{\Xi\backslash Z}\, h^{\#Z} \|\nabla_Z a\|,$$

with $\text{const}_{\Xi\backslash Z} := \min_{Y \in \mathcal{B}(\Xi\backslash Z)} |\det Y|$.

Proof. Since $a^h = m^h{*}a = h^{-s} b^h{*}a$, we conclude from (VI.13) that, for any $\mathrm{Z} \subseteq \Xi$,

$$\begin{aligned}\nabla_{h\mathrm{Z}} a^h &= h^{-s} \nabla_{h\mathrm{Z}} b^h * a \\ &= h^{\#\mathrm{Z}-s} \nabla_{\mathrm{Z}} b^h_{\Xi\backslash\mathrm{Z}} * a \\ &= h^{\#\mathrm{Z}-s} b^h_{\Xi\backslash\mathrm{Z}} * \nabla_{\mathrm{Z}} a \\ &= h^{n-s} \sum_{\alpha \in \square_h} \delta(\cdot - (\Xi\backslash\mathrm{Z})\alpha) * \nabla_{\mathrm{Z}} a,\end{aligned}$$

the last equality by (VI.3). Consequently

$$\text{(19)} \qquad \nabla_{h\mathrm{Z}} a^h = h^{n-s} \sum_{\alpha \in \square_h} \nabla_{\mathrm{Z}} a(\cdot - (\Xi\backslash\mathrm{Z})\alpha)$$

for any $\mathrm{Z} \subseteq \Xi$.

Now consider $\nabla_{h\mathrm{Z}} a^h(k)$ for $k \in h\mathbb{Z}^s$. Let Y be any basis in $\Xi\backslash\mathrm{Z}$, set $Y' := (\Xi\backslash\mathrm{Z})\backslash Y$, and set correspondingly

$$\mu(\beta,\gamma) := k - \mathrm{Y}'\beta - \mathrm{Y}\gamma, \quad \alpha =: (\beta,\gamma).$$

Then, the number of terms in the sum on the right hand side of (19) is bounded by

$$h^{-\#\mathrm{Y}'} \max_{\beta \in \square_h} \#\{\gamma \in \square_h : \mu(\beta,\gamma) \in \mathbb{Z}^s\}.$$

Since the map $\gamma \mapsto \mu(\beta,\gamma)$ is 1-1 and $\gamma \in \square_h \subset \square$,

$$\begin{aligned}\#\{\gamma \in \square_h : \mu(\beta,\gamma) \in \mathbb{Z}^s\} &\le \#\{j \in \mathbb{Z}^s : k - \mathrm{Y}'\beta - j \in \mathrm{Y}\square\} \\ &= \#\iota(k - \mathrm{Y}'\beta, \mathrm{Y}) = |\det \mathrm{Y}|,\end{aligned}$$

the last equality by (II.15). With this, it follows from (19) that

$$|\nabla_{h\mathrm{Z}} a^h(k)| \le h^{n-s} h^{-\#\mathrm{Y}'} |\det \mathrm{Y}| \|\nabla_{\mathrm{Z}} a\|.$$

□

Linear convergence of subdivision. Using (18)Theorem, the convergence of the refined coefficient sequence to the corresponding cardinal spline can be established under the rather weak assumption that

$$\text{(20)} \qquad \Xi\mathbb{Z}^n = \mathbb{Z}^s.$$

Equivalently formulated, this assumption states that any $y \in \mathbb{Z}^s$ can be written as a finite sum

$$y = \sum_{\nu} \eta_\nu$$

of (possibly repeated) vectors $\eta_\nu \in \Xi \cup -\Xi$. This implies that

$$\nabla_y a = \sum_\nu \nabla_{\eta_\nu} a(\cdot + \sum_{\mu<\nu} \eta_\mu). \tag{21}$$

For example, for $\Xi = \left[\begin{smallmatrix}3&3&1\\1&2&2\end{smallmatrix}\right]$ and $y = \mathbf{i}_1$, we have

$$\begin{aligned} a(1,0) - a(0,0) = a(1,0) &- a((1,0)+(3,2)) \\ &+ a(4,2) - a((4,2)-(3,1)) \\ &\quad + a(1,1) - a((1,1)+(3,2)) \\ &\qquad + a(4,3) - a((4,3)-(3,1)) \\ &\qquad\quad + a(1,2) - a(0,0), \end{aligned}$$

as is indicated in (22)Figure.

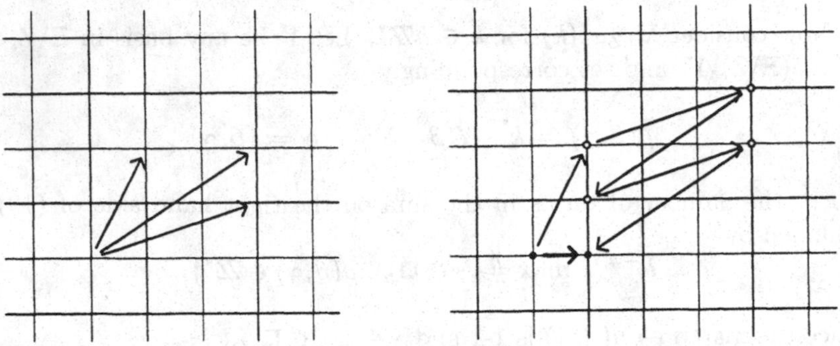

(22)Figure. Representation of the backward difference ∇_1 in terms of backward differences ∇_η, $\eta \in \Xi \cup -\Xi$, for $\Xi = \left[\begin{smallmatrix}3&3&1\\1&2&2\end{smallmatrix}\right]$.

We note (but make no use of the fact) that, by (VI.27)Corollary, (20) is equivalent to the condition that the greatest common divisor of all minors of Ξ of order s is 1.

(23)Theorem. *If $\Xi\mathbb{Z}^n = \mathbb{Z}^s$ and $M = M_\Xi$ is continuous, then*

$$|a^h(k) - (M*a)(x)| \le \text{const}'_\Xi \, h \max_{\xi\in\Xi} \|\nabla_\xi a\|$$

for $x - k \in h\Xi\mathbf{L}$.

Proof. Since M is continuous, $\Xi\backslash\xi$ spans for any $\xi \in \Xi$. This allows us to conclude from (18)Theorem that

$$c_1 := \max_{\xi\in\Xi} \|\nabla_{h\xi} a^h\| \le h \max_{\xi\in\Xi} \text{const}_{\Xi\backslash\xi} \|\nabla_\xi a\|.$$

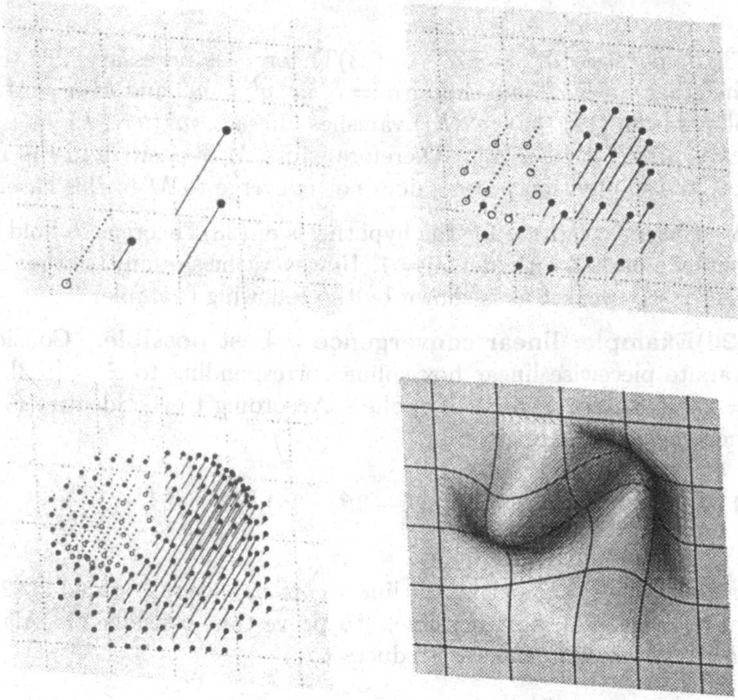

(24)Figure. The mesh functions a, $a^{1/2}$, $a^{1/4}$ as well as the corresponding cardinal spline obtained as limit of subdivision; the ZP element is the underlying box spline.

By definition of a^h, and since the shifts of M form a partition of unity,

(25) $$a^h(k) - M*a(x) = \sum_{\ell \in h\mathbb{Z}^s} M(\frac{x-\ell}{h})(a^h(k) - a^h(\ell)),$$

where the sum needs to be taken only over those ℓ for which $x - \ell \in h\Xi\square$. Since also $x - k \in h\Xi\square$,

$$|k-\ell| \le c_2 h, \quad c_2 := \operatorname{diam}(\Xi\square),$$

for those ℓ. Applying a scaled version of (21) with $y := k - \ell$, it follows that

$$|a^h(k) - a^h(\ell)| \le \sum_\nu \|\nabla_{h\xi_\nu} a^h(k + hj_\nu)\| \le c_3 c_1,$$

where c_3 is an upper bound for the number of terms in the sum, i.e., boundable in terms of c_2 and the number of terms in (21) when y is any one of the $\mathbf{i}_j$. This shows that the relevant coefficients in (25) are $O(h)$, uniformly in k and ℓ, and yields the desired estimate with

$$\text{const}'_\Xi := c_3 \max_\xi \text{const}_{\Xi\backslash\xi}.$$

□

The hypothesis $\mathbb{Z}^s = \Xi\mathbb{Z}^n$ of (23)Theorem is necessary. To see this, assume that $j \notin \Xi\mathbb{Z}^n$ and choose $a := \delta$, i.e. $a^h = m^h$ and $M{*}a = M$. Then it follows from (11) that $m^h(hj)$ vanishes. In fact, $m^h(hj+k)$ vanishes for any $k \in h\Xi\mathbb{Z}^n$ if $j \notin \Xi\mathbb{Z}^n$. Therefore, since M is positive in the interior of $\Xi\square$, the subdivision process does not converge to M in this case.

A sufficient condition for the hypothesis of (23)Theorem to hold is that Ξ contain a basis Z with $|\det \mathrm{Z}| = 1$. However, the assumption that $\Xi\mathbb{Z}^n = \mathbb{Z}^s$ is slightly weaker as is shown by the following example.

(26)Example: linear convergence is best possible. Consider the univariate piecewise linear box spline corresponding to $\Xi = [2\ 3]$. Since $\mathbb{Z} = 2\mathbb{Z} + 3\mathbb{Z}$, (23)Theorem applies. According to the identity (12), the refined coefficients are given by

(27) $$a^h(k) = h \sum_\alpha a(k - 2\beta - 3\gamma), \quad \alpha = (\beta, \gamma).$$

This example also shows that the linear convergence rate stated in (23)Theorem is, in general, best possible. To prove this, consider (16)Algorithm for $h = 1/2$ which in this case reduces to

$$\text{(i)}\ c := 2a^{h'}(\cdot h')$$

$$\text{(ii)}'\ c \longleftarrow (c + c(\cdot - 1))/2$$

$$c \longleftarrow (c + c(\cdot - 3/2))/2$$

$$a^{h'/2} := c(\cdot/h')$$

Combining these computations yields

$$a^{h'/2} = \left(a^{h'} + a^{h'}(\cdot - h') + a^{h'}(\cdot - 3h'/2) + a^{h'}(\cdot - 5h'/2)\right)/2.$$

For example, one computes

(28) $$\begin{aligned} 2a^{1/2}(2) &= a^1(1) + a^1(2) \\ 2a^{1/2}(5/2) &= a^1(0) + a^1(1) \\ 2a^{1/2}(3) &= a^1(2) + a^1(3) \\ 2a^{1/2}(7/2) &= a^1(1) + a^1(2), \end{aligned}$$

recalling our convention that $a^1(x) = 0$ if x is not an integer. With $c^{h'}$ denoting the vector of the four consecutive coefficients

$$\left[a^{h'}(4-4h'),\ a^{h'}(4-3h'),\ a^{h'}(4-2h'),\ a^{h'}(4-h')\right],$$

(28) is a special case of the identity

$$c^{h'/2} = \begin{bmatrix} 0 & 1/2 & 1/2 & 0 \\ 1/2 & 1/2 & 0 & 0 \\ 0 & 0 & 1/2 & 1/2 \\ 0 & 1/2 & 1/2 & 0 \end{bmatrix} c^{h'}.$$

One verifies for the particular choice $a^1 := \delta(\cdot - 1)$ (i.e. $c^1 = [0 \quad 1 \quad 0 \quad 0]'$) that

$$c^{h'} = \begin{bmatrix} 1/3 \\ 1/3 \\ 1/3 \\ 1/3 \end{bmatrix} + h' \begin{bmatrix} -1/3 \\ 2/3 \\ -1/3 \\ -1/3 \end{bmatrix}$$

for $h' = 2^{-2\nu}$, i.e. an even number of subdivision steps. Therefore, since $\max_x |M_\Xi(x)| = 1/3$, the sequence $a^{h'}$ can converge at most linearly.

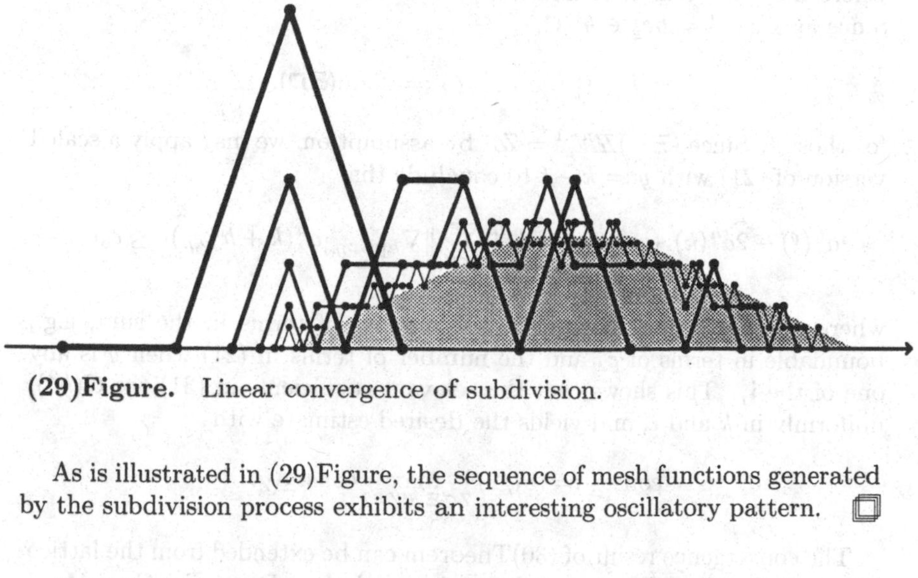

(29)Figure. Linear convergence of subdivision.

As is illustrated in (29)Figure, the sequence of mesh functions generated by the subdivision process exhibits an interesting oscillatory pattern. □

Quadratic convergence of subdivision. By slightly strengthening the hypothesis of (23)Theorem, a faster rate of convergence can be demonstrated. Note that the additional assumption $\Pi_1 \subseteq D(\Xi)$ is not sufficient for faster convergence since this condition is trivially satisfied for the preceding example.

(30)Theorem. *If* $(\Xi\backslash\xi)\mathbb{Z}^{n-1} = \mathbb{Z}^s$ *for all* $\xi \in \Xi$, *and* $M = M_\Xi$ *is continuously differentiable, then*

$$|a^h(k) - (M * a)(x)| \leq \text{const}''_\Xi \; h^2 \max_{\xi,\eta\in\Xi} \|\nabla_\xi \nabla_\eta a\|$$

for $x = k + hc_{\Xi}$, *with* $c_{\Xi} = \sum_{\xi\in\Xi} \xi/2$ *denoting the center of the support of* M.

Proof. The assumption that M is continuously differentiable ensures that $\Xi\backslash Z$ spans for every $Z \in \Xi$ with $\#Z = 2$. This allows us to conclude from (18)Theorem that

$$c_1 := \max_{Z\subset\Xi,\#Z=2} \|\nabla_{hZ} a^h\| \le h^2 \max_{Z\subset\Xi,\#Z=2} \text{const}_{\Xi\backslash Z} \|\nabla_Z a\|.$$

Further, since M is centrally symmetric around c_{Ξ}, we have $M((x-\ell)/h) = M((x+\ell-2k)/h)$ for our particular x and any ℓ. Consequently, (25) implies that

$$(31)\ \ a^h(k) - M{*}a(x) = \sum_{\ell\in h\mathbb{Z}^s} M(\frac{x-\ell}{h})(-1/2)(a^h(\ell) - 2a^h(k) + a^h(2k-\ell)),$$

where the sum needs to be taken only over those ℓ for which $x - \ell \in h\Xi\square$. Since also $x - k = hc_{\Xi} \in h\Xi\square$,

$$|k-\ell| \le c_2 h, \quad c_2 := \text{diam}(\Xi\sqcap),$$

for those ℓ. Since $(\Xi\backslash\xi)\mathbb{Z}^{n-1} = \mathbb{Z}^s$ by assumption, we may apply a scaled version of (21) with $y := k - \ell$ to conclude that

$$|a^h(\ell) - 2a^h(k) + a^h(2k-\ell)| \le \sum_{\nu,\mu} \|\nabla_{h[\eta_\nu,\eta_\mu]} a^h(k + hj_{\nu,\mu})\| \le c_3 c_1,$$

where c_3 is an upper bound for the number of terms in the sum, e.g., boundable in terms of c_2 and the number of terms in (21) when y is any one of the $\mathbf{i}_j$. This shows that the relevant coefficients in (31) are $O(h^2)$, uniformly in k and ℓ, and yields the desired estimate with

$$\text{const}''_{\Xi} := (c_3/2) \max_{Z\subset\Xi,\#Z=2} \text{const}_{\Xi\backslash Z}.$$

□

The convergence result of (30)Theorem can be extended from the lattice $h(\mathbb{Z}^s + c_{\Xi})$ to all of $\mathbb{R}^s$ via multilinear interpolation. Denote by $C := M_{\mathbb{1}\cup\mathbb{1}}$ the piecewise multilinear box spline, centered at the origin, i.e.,

$$C(x) = \prod_{\nu=1,\dots,s} M^c_{[1\,1]}(x(\nu))$$

where $M^c_{[1\,1]}$ is the univariate piecewise linear hat function which interpolates 1 at 0 and vanishes outside of the interval $(-1\,..\,1)$. Then, the multilinear interpolant corresponding to the coefficients a^h is given by

$$\mathcal{C}(\cdot, a^h) := (C^h * a^h)(\cdot - hc_{\Xi}).$$

This piecewise multilinear box spline is called the **control net** of the cardinal spline. Since $M^c_{[1\,1]|} = \delta$, it interpolates the coefficients a^h on the lattice $hc_\Xi + h\mathbb{Z}^s$, i.e.,

$$\mathcal{C}(k + hc_\Xi, a^h) = a^h(k) \qquad \forall k \in h\mathbb{Z}^s.$$

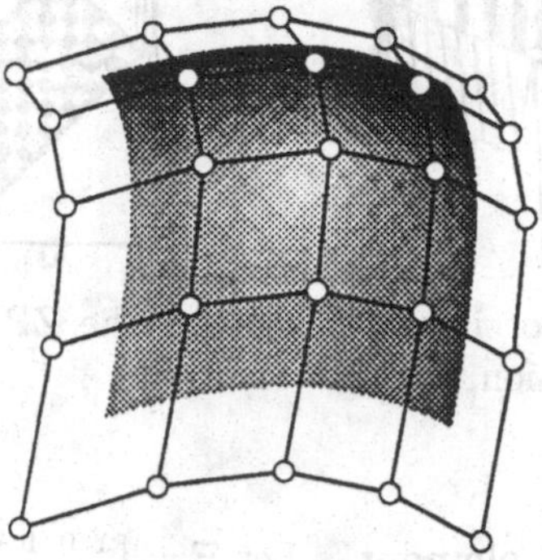

(32)Figure. Bilinear control polygon and corresponding cardinal spline.

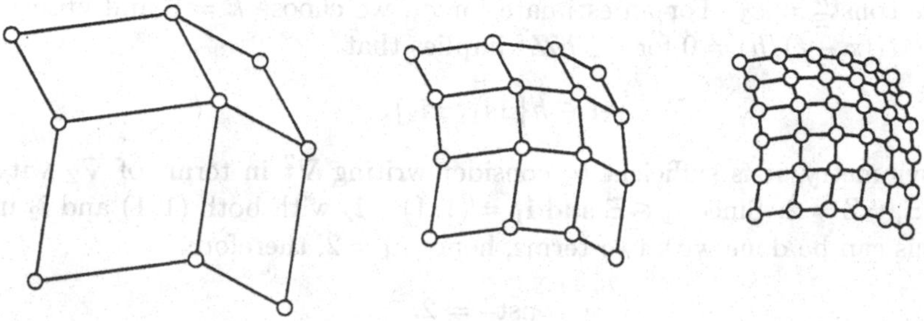

(33)Figure. Convergence of control nets generated by subdivision.

(34)Corollary. *Under the assumptions of (30)Theorem,*

$$|\mathcal{C}(x, a^h) - (M_\Xi * a)(x)| = O(h^2).$$

Approximation via the refinement of the control net yields a fast algorithm for graphic display of cardinal splines. For this, (16)Algorithm is applied repeatedly with $h = 1/2$ until successive control nets are sufficiently close. The number of necessary subdivision steps can be decided *a priori* by using an appropriate upper bound for const''_Ξ in (30)Theorem. We illustrate this for quadratic box splines.

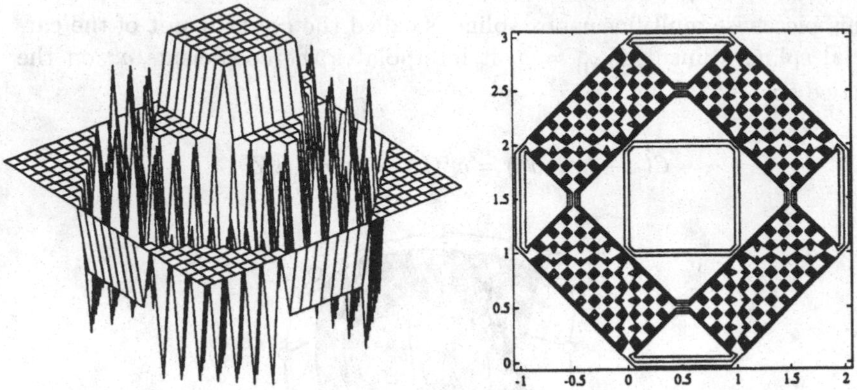

(35)Figure. The error in the values for the ZP element obtained by subdivision, with $h = 1/16$.

Example: the ZP element. Let $\Xi = \left[\begin{smallmatrix}1 & 0 & 1 & -1\\ 0 & 1 & 1 & 1\end{smallmatrix}\right]$. Then

$$\max_{\#Z=2} \text{const}_{\Xi\backslash Z} = 2,$$

hence $\text{const}''_{\Xi} = c_3$. For an estimate for c_3, we choose $k = 0$ and observe that $M((x-\ell)/h) \neq 0$ for $\ell \in h\mathbb{Z}^2$ implies that

$$\ell \in h\{\pm\mathbf{i}_1, \pm\mathbf{i}_2\}.$$

By symmetry, it is sufficient to consider writing ∇_1^2 in terms of ∇_Z with $Z \subset \Xi$, $\#Z = 2$. Since $\mathbf{i}_1 \in \Xi$ and $\mathbf{i}_1 = (1,1) - \mathbf{i}_2$ with both $(1,1)$ and $\mathbf{i}_2$ in Ξ, this can be done with two terms, hence $c_3 = 2$, therefore

$$\text{const}''_{\Xi} = 2.$$

In fact, calculations with $h = 2^{-j}$, $j = 2,3,4,5$ show the maximum error to be $(h/2)^2$. Moreover, the error shows rather remarkable regularity which suggests that extrapolation to the limit should be an effective tool for improving the accuracy of values obtained by subdivision. □

The subdivision algorithms can be extended to **box spline surfaces**, i.e., surfaces with a parametrization of the form

$$(36) \qquad x \mapsto \sum_j M_\Xi(x-j)a(j), \quad x \in \mathbb{R}^2,$$

with coefficients $a(j) \in \mathbb{R}^3$ and x restricted to an appropriate parameter domain. The subdivision algorithm is applied simultaneously in each component. The whole process is illustrated in (37)Figure, in which we start off with a control net consisting of just 4×4 control points. The result of the first subdivision, also shown, increases this to a control net of 6×6 control points. The surface described by the original control net is also shown. Actually, for the sake of uniformity (which permits use of parallelism in the calculations), the subdivision process operates on a biinfinite control net which is obtained from the given one by choosing the additional control points to be 0. This means that the process needs to be carried out only 'near' the original control points. On the other hand, it means that one has to keep track of which control points in the refined control nets generated are relevant for the surface piece fully described by the original control net. Only these should be retained at the end.

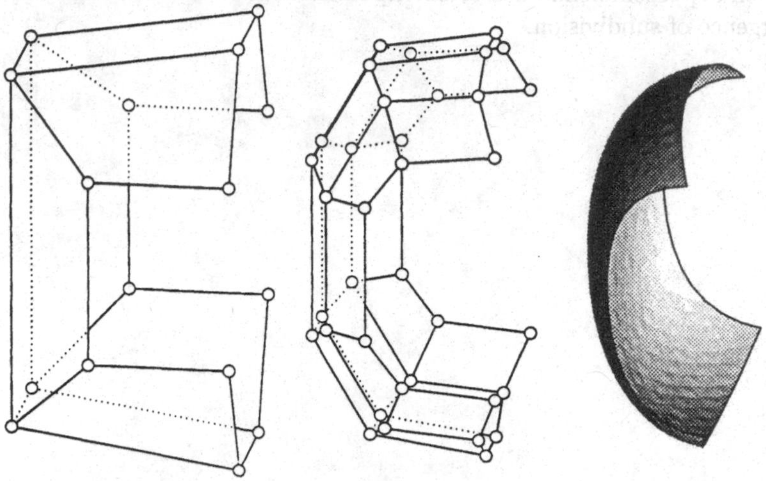

(37)Figure. Control net converging to a box spline surface.

Notes. The idea of *subdivision algorithms* originated in CAGD, with Chaikin's paper [Chaikin'74] and Riesenfeld's proof [Riesenfeld'75] that Chaikin's algorithm converges to a quadratic cardinal spline. (However, the study of limits of such subdivision algorithms, in the guise of corner-cutting algorithms, goes back at least to [de Rham'47].) Boehm [Boehm'83a] carried this idea to the triangular splines of (Frederickson and) Sabin [Sabin'77], using masks to compute the refined control nets. Cohen, Lyche and Riesenfeld [Cohen, Lyche, Riesenfeld'84] obtained masks for arbitrary box splines by subdividing □, leading to (15)Algorithm. Dahmen and Micchelli [Dahmen, Micchelli'84a,'85b] obtained such algorithms by algebraic means.

(18)Theorem on *subdivision as discrete smoothing* is new.

Linear convergence of the subdivision algorithm was proved in [Dahmen, Micchelli'84a], using a connection to numerical quadrature, but the particular (26)Example showing that this is, in general, best possible is new. The 'optimal' order, h^2, was proved under certain assumptions on Ξ in [Dahmen, Dyn, Levin'85]. Our argument for (30)Theorem relies instead on (18)Theorem, and this allows the assumptions on Ξ made in (30)Theorem to be considerably weaker than in [Dahmen, Dyn, Levin'85]. On the other hand, [Dahmen, Dyn, Levin'85] contains examples in which the convergence is locally faster than $O(h^2)$ (without having convergence occur in finitely many steps).

[Prautzsch'85] employs some refined averaging (related to higher-order quadrature) to obtain higher-order convergence rates to the spline, while [Dahmen'87] recommends use of extrapolation for the acceleration of the convergence of subdivision.

References

J. F. Adams (1962): Vector fields on spheres, Ann. Math. **75**, 603–632.

E. Arge, M. Dæhlen, and A. Tveito (1991): Box spline interpolation: a computational study, preprint.

R. B. Barrar, and H. L. Loeb (1984): Some remarks on the exact controlled approximation order of bivariate splines on a diagonal mesh, J. Approx. Theory **42**, 257–265.

G. Battle (1987): A block spin construction of ondelettes, Part I: Lemarie Functions, Comm. Math. Phys. **110**, 601–615.

A. Ben-Artzi, and A. Ron (1988): Translates of exponential box splines and their related spaces, Trans. Amer. Math. Soc. **309**, 683–710.

A. Ben-Artzi, and A. Ron (1990): On the integer translates of a compactly supported function: dual bases and linear projectors, SIAM J. Math. Anal. **21**, 1550–1562.

Peter G. Binev (1988): Error estimate for box spline interpolation, in *Constructive Theory of Functions '87*, B. Sendov, P. Petrushev, K. Ivanov, and R. Maleev (eds.), Bulgarian Academy of Sciences, Sofia, 50–55.

Peter G. Binev, and K. Jetter (1991a): Euler splines from 3-directional box splines, in *Constructive Theory of Functions '91*, K. Ivanov et al. (eds.), Bulgarian Academy of Sciences, Sofia, 1–8.

P.G. Binev, and K. Jetter (1992a): Cardinal interpolation with shifted 3-directional box splines, Proc. Roy. Soc. Edinburgh Sect. A **122A**, 205–220.

P.G. Binev, and K. Jetter (1992b): Estimating the condition number for multivariate interpolation problems, in *Numerical Methods in Approximation Theory*, D. Braess, L.L. Schumaker (eds.), Birkhäuser, Basel, 39–50.

W. Boehm (1983a): The de Boor algorithm for triangular splines, in *Surfaces in Computer Aided Geometric Design*, R. E. Barnhill and W. Boehm (eds.), North Holland, Amsterdam, 109–120.

W. Boehm (1983b): Generating the Bézier points of triangular splines, in *Surfaces in Computer Aided Geometric Design*, R. E. Barnhill and W. Boehm (eds.), North Holland, Amsterdam, 77–92.

W. Boehm (1983d): Subdividing multivariate splines, Computer-Aided Design **15**, 345–352.

W. Boehm (1984): Calculating with box splines, Comput. Aided Geom. Design **1**, 149–162.

W. Boehm (1985a): Triangular spline algorithms, Comput. Aided Geom. Design **2**, 61–67.

W. Boehm (1985c): Multivariate spline algorithms, Computer-Aided Design **17**, 103–105.

W. Boehm (1986a): Multivariate spline algorithms, in *The Mathematics of Surfaces*, J. Gregory (ed.), Clarendon Press, Oxford, 197–215.

W. Boehm (1986b): Multivariate spline methods in CAGD, Comput. Aided Geom. Design **18**, 102–104.

W. Boehm, G. Farin, and J. Kahmann (1984): A survey of curve and surface methods in CAGD, Comput. Aided Geom. Design **1**, 1–60.

W. Boehm, H. Prautzsch, and P. Arner (1987): On triangular splines, Constr. Approx. **3**, 157–167.

J. Bohne, A. Dress, and S. Fischer (1989): A simple proof for de Bruin's dualization principle, in *Some elementary proofs for some elementary facts in algebra and geometry* by Andreas W. M. Dress, Preprint 89-027, SFB 343 'Diskrete Strukturen der Mathematik', Universität Bielefeld, Germany.

C. de Boor (1968b): On uniform approximation by splines, J. Approx. Theory **1**, 219–235.

C. de Boor (1976h): Splines as linear combinations of B-splines, a survey, in *Approximation Theory II*, G. G. Lorentz, C. K. Chui, and L. L. Schumaker (eds.), Academic Press, New York, 1–47.

C. de Boor (1978b): *A Practical Guide to Splines*, Springer Verlag, New York.

C. de Boor (1987c): The polynomials in the linear span of integer translates of a compactly supported function, Constr. Approx. **3**, 199–208.

C. de Boor (1988b): What is a multivariate spline?, in *Proc. First Intern. Conf. Industr. Applied Math., Paris 1987*, J. McKenna and R. Temam eds., 90–101.

C. de Boor (1990c): Quasiinterpolants and approximation power of multivariate splines, in *Computation of Curves and Surfaces*, W. Dahmen, M. Gasca and C. Micchelli (eds.), Kluwer, 313–345.

C. de Boor (1993): On the evaluation of box splines, in *Algorithms for Approximation III*, Maurice G. Cox (ed.), xxx (xxx), xxx–xxx.

C. de Boor, and R. DeVore (1983): Approximation by smooth multivariate splines, Trans. Amer. Math. Soc. **276**, 775–788.

C. de Boor, R. DeVore, and A. Ron (1992a): Approximation from shift-invariant subspaces of $L_2(\mathbb{R}^d)$, CMS-TSR University of Wisconsin-Madison **92-2**.

C. de Boor, R. DeVore, and A. Ron (1993): On the construction of multivariate (pre)wavelets, Constr. Approx. **xx**, xxx–xxx.

C. de Boor, N. Dyn, and A. Ron (1991): On two polynomial spaces associated with a box spline, Pacific J. Math. **147**, 249–267.

C. de Boor, and K. Höllig (1982a): Recurrence relations for multivariate B-splines, Proc. Amer. Math. Soc. **85**, 397–400.

C. de Boor, and K. Höllig (1982b): B-splines from parallelepipeds, J. Analyse Math. **42**, 99–115.

C. de Boor, and K. Höllig (1983a): Approximation order from bivariate C^1-cubics: a counterexample, Proc. Amer. Math. Soc. **87**, 649–655.

C. de Boor, and K. Höllig (1983b): Bivariate box splines and smooth pp functions on a three direction mesh, J. Comput. Appl. Math. **9**, 13–28.

C. de Boor, and K. Höllig (1987a): Minimal support for bivariate splines, Approx. Theory Appl. **3**, 11–23.

C. de Boor, and K. Höllig (1991): Box-spline tilings, Amer. Math. Monthly **98**, 793–802.

C. de Boor, K. Höllig, and S. Riemenschneider (1983): Bivariate cardinal interpolation, in *Approximation Theory IV*, C. Chui, L. Schumaker, and J. Ward (eds.), Academic Press, New York, 359–363.

C. de Boor, K. Höllig, and S. Riemenschneider (1985a): The limits of multivariate cardinal splines, in *Multivariate Approximation Theory III*, W. Schempp and K. Zeller (eds.), Birkhäuser, Basel, 47–50.

C. de Boor, K. Höllig, and S. Riemenschneider (1985b): Bivariate cardinal interpolation by splines on a three-direction mesh, Illinois J. Math. **29**, 533–566.

C. de Boor, K. Höllig, and S. Riemenschneider (1985c): Convergence of bivariate cardinal interpolation, Constr. Approx. **1**, 183–193.

C. de Boor, K. Höllig, and S. Riemenschneider (1986b): Convergence of cardinal series, Proc. Amer. Math. Soc. **98**, 457–460.

C. de Boor, K. Höllig, and S. Riemenschneider (1987): Some qualitative properties of bivariate Euler-Frobenius polynomials, J. Approx. Theory **50**, 8–17.

C. de Boor, K. Höllig, and S. Riemenschneider (1989): Fundamental solutions for multivariate difference equations, Amer. J. Math. **111**, 403–415.

C. de Boor, and A. Ron (1991a): On polynomial ideals of finite codimension with applications to box spline theory, J. Math. Anal. Appl. **158**, 168–193.

C. de Boor, and A. Ron (1992b): Computational aspects of polynomial interpolation in several variables, Math. Comp. **58**, 705–727.

C. de Boor, and A. Ron (1992c): The exponentials in the span of the multiinteger translates of a compactly supported function: quasi-interpolation and approximation order, J. London Math. Soc. (2) **45**, 519–535.

C. de Boor, and A. Ron (1992d): Fourier analysis of the approximation power of principal shift-invariant spaces, Constr. Approx. **8**, 427–462.

A. Cavaretta, and C. A. Micchelli (1989a): Subdivision algorithms, in *Mathematical Methods in Computer Aided Geometric Design*, T. Lyche and L. Schumaker (eds.), Academic Press, N. Y., 115–153.

G. M. Chaikin (1974): An algorithm for high speed curve generation, Computer Graphics and Image Processing **3**, 346–349.

C. K. Chui (1988): *Multivariate Splines*, CBMS-NSF Series Appl. Math. Vol.54, SIAM Publications, Philadelphia.

C. K. Chui, and H. Diamond (1987): A natural formulation of quasi-interpolation by multivariate splines, Proc. Amer. Math. Soc. **99**, 643–646.

C. K. Chui, H. Diamond, and L. A. Raphael (1988a): Convexity-preserving quasi-interpolation and interpolation by box spline surfaces, in *Transactions of the Fifth Army Conference on Applied Mathematics and Computing*, U.S. Army Res. Office, Research Triangle Park, NC, 301–310.

C. K. Chui, H. Diamond, and L. A. Raphael (1988b): Interpolation by multivariate splines, Math. Comp. **51**, 203–218.

C. K. Chui, H. Diamond, and L. Raphael (1989): Shape-preserving quasi-interpolation and interpolation by box spline surfaces, J. Comput. Appl. Math. **25**, 169–198.

C. K. Chui, K. Jetter, and J. D. Ward (1987): Cardinal interpolation by multivariate splines, Math. Comp. **48**, 711-724.

C. K. Chui, and M. J. Lai (1987a): A multivariate analog of Marsden's identity and a quasi-interpolation scheme, Constr. Approx. **3**, 111–122.

C. K. Chui, and M. J. Lai (1987c): Computation of box splines and B-splines on triangulations of nonuniform rectangular partitions, Approx. Theory Appl. **3**, 37–62.

C. K. Chui, M. J. Lai, and S. R. Bowers (1988): An algorithm for generating B-nets and graphically displaying box-splines surfaces, CAT Report 181, Texas A&M University.

C. K. Chui, and A. Ron (1991): On the convolution of a box spline with a compactly supported distribution: linear independence for the integer translates, Canad. J. Math. **43**, 19–33.

C. K. Chui, J. Stöckler, and J. D. Ward (1989a): Bivariate cardinal interpolation with a shifted box-spline on a three-directional mesh, CAT Report 188, Texas A&M University.

C. K. Chui, J. Stöckler, and J. D. Ward (1989b): Cardinal interpolation with shifted box-splines, in *Approximation Theory VI*, C. Chui, L. Schumaker, and J. Ward (eds.), Academic Press, New York, 141–144.

C. K. Chui, J. Stöckler, and J. D. Ward (1991a): Invertibility of shifted box spline interpolation operators, SIAM J. Math. Anal. **22**, 543–553.

C. K. Chui, J. Stöckler, and J. D. Ward (1992a): A Faber series approach to cardinal interpolation, Math. Comp. **58**, 255–273.

C. K. Chui, J. Stöckler, and J. D. Ward (1992b): Compactly supported box spline wavelets, Approx. Theory Appl. **8**, 77–100.

C. K. Chui, J. Stöckler, and J. D. Ward (199xa): Singularity of cardinal interpolation with shifted box splines, CAT Report #217, Texas A&M University, to appear in J. Approx. Theory .

C. K. Chui, and R. H. Wang (1983a): Multivariate spline spaces, J. Math. Anal. Appl. **94**, 197–221.

C. K. Chui, and R. H. Wang (1984b): On a bivariate B-spline basis, Scientia Sinica **27**, 1129–1142.

C. K. Chui, and J. Z. Wang (1992a): On compactly supported spline wavelets and a duality principle, Trans. Amer. Math. Soc. **330**, 903–915.

C. K. Chui, and Jian Zhong Wang (1992b): A general framework for compactly supported splines and wavelets, J. Approx. Theory **71(3)**, 263–304.

P. G. Ciarlet (1988): *Introduction to Numerical Linear Algebra and Optimization*, Cambridge University Press.

E. Cohen, T. Lyche, and R. Riesenfeld (1980): Discrete B-splines and subdivision techniques in computer-aided geometric design and computer graphics, Computer Graphics Image Proc. **14**, 87–111.

E. Cohen, T. Lyche, and R. Riesenfeld (1984): Discrete box splines and refinement algorithms, Comput. Aided Geom. Design **1**, 131–148.

R. Courant (1943): Variational methods for the solution of problems in equilibrium and vibrations, Bull. Amer. Math. Soc. **49**, 1–23.

H. B. Curry, and I. J. Schoenberg (1966): On the Pólya frequency functions IV: the fundamental spline functions and their limits, J. Analyse Math. **17**, 71–107.

M. Dæhlen (1987): An example of bivariate interpolation with translates of C^0-quadratic box-splines on a three direction mesh, Comput. Aided Geom. Design **4**, 251–255.

M. Dæhlen (1989a): On the evaluation of box-splines, in *Mathematical Methods in Computer Aided Geometric Design*, T. Lyche and L. Schumaker (eds.), Academic Press, N. Y., 167–179.

M. Dæhlen (1989b): Box splines and applications of polynomial splines, dissertation, University of Oslo, Research Report 128. Inst. for Informatics.

M. Dæhlen (1992a): Modelling with box spline surfaces, in *Mathematical Methods in Computer Aided Geometric Design*, T. Lyche and L. Schumaker (eds.), Academic Press, N. Y., xxx–xxx.

M. Dæhlen, and T. Lyche (1988): Bivariate interpolation with quadratic box splines, Math. Comp. **51**, 219–230.

M. Dæhlen, and T. Lyche (1991): Box splines and applications, in *Geometric Modelling: Methods and Applications*, H. Hagen and D. Roller (eds.), Springer-Verlag, Berlin, 35–93.

M. Dæhlen, and V. Skytt (1989): Modelling non-rectangular surfaces using box splines, in *Mathematics of Surfaces III*, D. C. Handscomb (ed.), Clarendon Press, Oxford, 285–300.

W. Dahmen (1980): On multivariate B-splines, SIAM J. Numer. Anal. **17**, 179–190.

W. Dahmen (1986a): Subdivision algorithms converge quadratically, J. Comput. Appl. Math. **16**, 145–158.

W. Dahmen (1987): Subdivision algorithms - recent results, some extensions and further developments, in *Algorithms for the Approximation of Functions and Data*, J. C. Mason and M. G. Cox (eds.), Oxford Univ. Press, 21–49.

W. Dahmen (1990): A basis of certain spaces of multivariate polynomials and exponentials, in *Algorithms for Approximation II*, M. G. Cox and J. C. Mason (eds.), Chapman & Hall, London, 80–98.

W. Dahmen, A. Dress, and C. A. Micchelli (1990): On multivariate splines, matroids, and the Ext-functor, IBM, Research report RC 16192.

W. Dahmen, N. Dyn, and D. Levin (1985): On the convergence rates of subdivision algorithms for box spline surfaces, Constr. Approx. **1**, 305–322.

W. Dahmen, Rong-Qing Jia, and C. Micchelli (1989): Linear dependence of cube splines revisited, in *Approximation Theory VI*, C. Chui, L. Schumaker, and J. Ward (eds.), Academic Press, New York, 161–164.

W. Dahmen, Rong-Qing Jia, and C. A. Micchelli (1991a): On linear dependence relations for integer translates of compactly supported distributions, Math. Nachrichten **151**, 303–310.

W. Dahmen, and C. A. Micchelli (1983b): Translates of multivariate splines, Linear Algebra Appl. **52**, 217–234.

W. Dahmen, and C. A. Micchelli (1983d): Multivariate splines—A new constructive approach, in *Surfaces in Computer Aided Geometric Design*, R. E. Barnhill and W. Boehm (eds.), North Holland, Amsterdam, 191–215.

W. Dahmen, and C. A. Micchelli (1984a): Subdivision algorithms for the generation of box-spline surfaces, Comput. Aided Geom. Design **1**, 115–129.

W. Dahmen, and C. A. Micchelli (1984b): On the approximation order from certain multivariate spline spaces, J. Austral. Math. Society, Ser. B **26**, 233–246.

W. Dahmen, and C. A. Micchelli (1984c): On the optimal approximation rates for criss-cross finite element spaces, J. Comput. Appl. Math. **10**, 255–273.

W. Dahmen, and C. A. Micchelli (1984d): Some results on box splines, Bull. Amer. Math. Soc. **11**, 147–150.

W. Dahmen, and C. A. Micchelli (1984e): On the multivariate Euler-Frobenius polynomials, in *Constructive Theory of Functions '84*, B. Sendov, P. Petrushev, R. Maleev, and S. Tashev (eds.), Bulgarian Academy of Sciences, Sofia, 237–243.

W. Dahmen, and C. A. Micchelli (1984f): Recent progress in multivariate splines, in *Approximation Theory IV*, C. Chui, L. Schumaker, and J. Ward (eds.), Academic Press, New York, 27–121.

W. Dahmen, and C. A. Micchelli (1985a): Combinatorial aspects of multivariate splines, in *Multivariate Approximation Theory III*, W. Schempp and K. Zeller (eds.), Birkhäuser, Basel, 130–137.

W. Dahmen, and C. A. Micchelli (1985b): Line average algorithm: a method for the computer generation of smooth surfaces, Comput. Aided Geom. Design **2**, 77–85.

W. Dahmen, and C. A. Micchelli (1985c): On the solution of certain systems of partial difference equations and linear independence of translates of box splines, Trans. Amer. Math. Soc. **292**, 305–320.

W. Dahmen, and C. A. Micchelli (1985d): On the local linear independence of translates of a box spline, Studia Math. **82**, 243–262.

W. Dahmen, and C. A. Micchelli (1986c): Statistical encounters with B-splines, Contemporary Math. **59**, 17–48.

W. Dahmen, and C. A. Micchelli (1986d): On the piecewise structure of discrete box splines, Comput. Aided Geom. Design **3**, 185–191.

W. Dahmen, and C. A. Micchelli (1987a): Algebraic properties of discrete box splines, Constr. Approx. **3**, 209–221.

W. Dahmen, and C. A. Micchelli (1987b): On the theory and application of exponential splines, in *Topics in Multivariate Approximation*, C. K. Chui, L. L. Schumaker, and F. Utreras (eds.), Academic Press, New York, 37–46.

W. Dahmen, and C. A. Micchelli (1988c): The number of solutions to linear Diophantine equations and multivariate splines, Trans. Amer. Math. Soc. **308**, 509–532.

W. Dahmen, and C. A. Micchelli (1988d): Convexity of multivariate Bernstein polynomials and box spline surfaces, Studia Math. **23**, 265–287.

W. Dahmen, and C. A. Micchelli (1989a): On multivariate E-splines, Advances in Math. **76**, 33–93.

W. Dahmen, and C. A. Micchelli (1990a): Local dimension of piecewise polynomial spaces, syzygies, and solutions of systems of partial differential equations, Mathem. Nachr. **148**, 117–136.

W. Dahmen, and C. A. Micchelli (1990b): Convexity and Bernstein polynomials on k-simploids, Acta Mathematicae Applicatae Sinica **6**, 50–66.

W. Dahmen, C. A. Micchelli, and T. N. T. Goodman (1989): Local spline schemes in one and several variables, in *Approximation and Optimization, Lecture Notes in Mathematics* **1354**, A. Gomez, F. Guerra, M.A. Jimenez, G. Lopez eds., Springer-Verlag, New York, 11–24.

N. Dyn, and A. Ron (1989): Periodic exponential box splines on a three direction mesh, J. Approx. Theory **56**, 287–296.

N. Dyn, and A. Ron (1990): Local approximation by certain spaces of multivariate exponential-polynomials, approximation order of exponential box splines and related interpolation problems, Trans. Amer. Math. Soc. **319**, 381–404.

R. Farwig (1985b): Multivariate truncated powers and B-splines with coalescent knots, SIAM J. Numer. Anal. **22**, 592–603.

G. Fix, and G. Strang (1969): Fourier analysis of the finite element method in Ritz-Galerkin theory, Studies in Appl. Math. **48**, 265–273.

R. Franke, and L. L. Schumaker (1987): A bibliography of multivariate approximation, in *Topics in Multivariate Approximation*, C. K. Chui, L. L. Schumaker, and F. Utreras (eds.), Academic Press, New York, 275–335.

P. O. Frederickson (1970): Triangular spline interpolation, Rpt. 6–70, Lakehead Univ.

P. O. Frederickson (1971b): Generalized triangular splines, Rpt. 7–71, Lakehead Univ.

T. N. T. Goodman (1985a): Shape preserving approximation by polyhedral splines, in *Multivariate Approximation Theory III*, W. Schempp and K. Zeller (eds.), Birkhäuser, Basel, 198–205.

T. N. T. Goodman (1989a): Shape preserving representations, in *Mathematical Methods in Computer Aided Geometric Design*, T. Lyche and L. Schumaker (eds.), Academic Press, N. Y., 333–351.

T. N. T. Goodman, S. L. Lee, and A. Sharma (1989): Approximation and interpolation by complex splines on the torus, Proc. Edinburgh Math. Soc.(Series II) **32**, 197–212.

T. N. T. Goodman, and A. A. Taani (1990): Cardinal interpolation by symmetric exponential box splines on a three-direction mesh, Proc. Edinburgh Math. Soc.(Series II) **33**, 251–264.

R. P. Gosselin (1963): On the L^p theory of cardinal series, Annals of Math. **78**, 567–581.

K. Gröchenig (1987): Analyse multi-échelles et bases d'ondelettes, C. R. Acad. Sci. Paris Sér. I Math. **305**, 13–15.

F. di Guglielmo (1970): Méthode des éléments finis: une famille d'approximations des espaces de Sobolev par les translates de *p* fonctions, Calcolo **7**, 185–233.

Zhu Rui Guo, and Rong-Qing Jia (1990): A B-net approach to the study of multivariate splines, Advances in Mathematics (China). Shuxue Jinzhan **19**, 189–198.

M. H. Gutknecht (1987): Attenuation factors in multivariate Fourier analysis, Numer. Math. **51**, 615–629.

A. A. Akopyan, and A. A. Saakyan (1988): A system of differential equations that is related to the polynomial class of translates of a box spline, Math. Notes **44**, 865–878.

A. A. Akopyan, and A. A. Saakyan (1989): A class of systems of partial differential equations, Izvestiya Akademii Nauk Armyanskoi SSR. Seriya Matematika **24**, 93–98.

J. R. Higgins (1985): Five short stories about the cardinal series, Bull. Amer. Math. Soc., New Series **12**, 45–89.

K. Höllig (1982a): Multivariate splines, SIAM J. Numer. Anal. **19**, 1013–1031.

K. Höllig (1982b): A remark on multivariate B-splines, J. Approx. Theory **33**, 119–125.

K. Höllig (1986b): Box splines, in *Approximation Theory V*, C. Chui, L. Schumaker, and J. Ward (eds.), Academic Press, New York, 71–95.

K. Höllig (1986c): Multivariate splines, in *Approximation Theory, Proc. Symp. Appl. Math.* **36**, C. de Boor (ed.), Amer.Math.Soc., Providence, 103–127.

K. Höllig (1989a): Box-spline surfaces, in *Mathematical Methods in Computer Aided Geometric Design*, T. Lyche and L. Schumaker (eds.), Academic Press, N. Y., 385–402.

K. Höllig, M. J. Marsden, and S. D. Riemenschneider (1989): Bivariate cardinal interpolation on the 3-direction mesh: ℓ^p-data, Rocky Mountain J. Math. **19**, 189–198.

L. Hörmander (1958): On the division of distributions by polynomials, Arkiv for Matematik **54**, 555–568.

A. Hurwitz (1923): Über die Komposition der quadratischen Formen, Math. Ann. **88**, 1–25.

T. Jensen (1987): Assembling triangular and rectangular patches and multivariate splines, in *Geometric Modeling: Algorithms and New Trends*, G. E. Farin (ed.), SIAM Publications, Philadelphia, 203–220.

K. Jetter (1987a): A short survey on cardinal interpolation by box splines, in *Topics in Multivariate Approximation*, C. K. Chui, L. L. Schumaker, and F. Utreras (eds.), Academic Press, New York, 125–139.

K. Jetter (1992a): Multivariate approximation: A view from cardinal interpolation, in *Approximation Theory VII*, E. W. Cheney, C. Chui, and L. Schumaker (eds.), Academic Press, New York, 131–161.

K. Jetter, and P. Koch (1989): Methoden der Fourier-Transformation bei der kardinalen Interpolation periodischer Daten, in *Multivariate Approximation Theory IV*, C. Chui, W. Schempp, and K. Zeller (eds.), Birkhäuser Verlag, Basel, 201–208.

K. Jetter, and S. Riemenschneider (1986a): Cardinal interpolation with box splines on submodules of $\mathbb{Z}^d$, in *Approximation Theory V*, C. Chui, L. Schumaker, and J. Ward (eds.), Academic Press, New York, 403–406.

K. Jetter, and S. Riemenschneider (1987): Cardinal interpolation, submodules, and the 4-direction mesh, Constr. Approx. **3**, 169–188.

K. Jetter, and J. Stöckler (1991a): Algorithms for cardinal interpolation using box splines and radial basis functions, Numer. Math. **60**, 97–114.

Rong-Qing Jia (1983d): Approximation by smooth bivariate splines on a three-direction mesh, in *Approximation Theory IV*, C. Chui, L. Schumaker, and J. Ward (eds.), Academic Press, New York, 539–545.

Rong-Qing Jia (1984b): Linear independence of translates of a box spline, J. Approx. Theory **40**, 158–160.

Rong-Qing Jia (1985): Local linear independence of the translates of a box splines, Constr. Approx. **1**, 175–182.

Rong-Qing Jia (1986a): Approximation order from certain spaces of smooth bivariate splines on a three-direction mesh, Trans. Amer. Math. Soc. **295**, 199–212.

Rong-Qing Jia (1986b): A counterexample to a result concerning controlled approximation, Proc. Amer. Math. Soc. **97**, 647–654.

Rong-Qing Jia (1987b): Recent progress in the study of box splines, Appl. Math. (a journal of Chinese Univ.s) **3**, 330–342.

Rong-Qing Jia (1988b): Local approximation order of box splines, Scientia Sinica **31**, 274–285.

Rong-Qing Jia (1989c): Dual bases associated with box splines, in *Multivariate Approximation Theory IV*, C. Chui, W. Schempp, and K. Zeller (eds.), Birkhäuser Verlag, Basel, 209–216.

Rong-Qing Jia (1990a): Approximation order of translation invariant subspaces of functions, in *Approximation Theory VI*, C. Chui, L. Schumaker, and J. Ward (eds.), Academic Press, New York, 349–352.

Rong-Qing Jia (1990e): Subspaces invariant under translation and the dual bases for box splines, Chinese Ann.Math. **11A**, 733–743.

Rong-Qing Jia (1991a): Multivariate discrete splines and linear diophantine equations, ms.

Rong-Qing Jia (1991b): The conjecture of Stanley for symmetric magic squares, preprint.

Rong-Qing Jia (1993a): A dual basis for the integer translates of an exponential box spline, Rocky Mountain J. Math. **23**, 223–242.

Rong-Qing Jia (199x): Multivariate discrete splines and linear diophantine equations, Trans. Amer. Math. Soc. **xx**, xxx–xxx.

Rong-Qing Jia, and C. A. Micchelli (1991a): Using the refinement equation for the construction of pre-wavelets II: power of two, in *Approximation Theory VI*, C. Chui, L. Schumaker, and J. Ward (eds.), Academic Press, New York, 209–246.

Rong-Qing Jia, and C. A. Micchelli (1992a): Using the refinement equation for the construction of pre-wavelets V: extensibility of trigonometric polynomials, Computing **48**, 61–72.

Rong-Qing Jia, and Zuowei Shen (199x): Multiresolution and wavelets, preprint.

Rong-Qing Jia, and N. Sivakumar (1990): On the linear independence of integer translates of box splines with rational directions, Linear Algebra Appl. **135**, 19–31.

S. Karlin, C. A. Micchelli, and Y. Rinott (1986): Multivariate splines: a probabilistic perspective, J. Multivariate Anal. **20**, 69–90.

P. Kochevar (1984): An application of multivariate B-splines to computer-aided geometric design, Rocky Mountain J. Math. **14**, 159–175.

Jan Krzysztof Kowalski (1990a): Application of box splines to the approximation of Sobolev spaces, J. Approx. Theory **61**, 53–73.

Jan Krzysztof Kowalski (1990b): A method of approximation of Besov spaces, Studia Math. **96**, 183–193.

Ming Jun Lai (1989b): A remark on translates of a box spline, Approx. Theory Appl. **5**, 97–104.

Ming Jun Lai (1992): Fortran subroutines for B-nets of box splines on three- and four-directional meshes, Numer. Algorithms **2**, 33–38.

J. M. Lane, and R. F. Riesenfeld (1980): A theoretical development for the computer generation and display of piecewise polynomial surfaces, IEEE Trans. Pattern Anal. Mach. Intellig. **2**, 35–45.

D. Lee (1986): A note on bivariate box splines on a k-direction mesh, J. Comput. Appl. Math. **15**, 117–122.

Junjiang Lei, and Rong-Qing Jia (1991): Approximation by piecewise exponentials, SIAM J. Math. Anal. **22**, 1776–1789.

S. Łojasiewicz (1959): Sur le problème de la division, Studia Math. **18**, 87–136.

R. A. H. Lorentz, and W. R. Madych (1992a): Wavelets and generalized box splines, Appl. Anal. **44**, 51–76.

S. G. Mallat (1989a): Multiresolution approximations and wavelet orthonormal bases of $L^2(\mathbb{R})$, Trans. Amer. Math. Soc. **315**, 69–87.

W. H. McCrea, and F. J. W. Whipple (1940): Random paths in two and three dimensions, Proc. Roy. Soc. Edinburgh Sect. A **60**, 281–298.

Y. Meyer (1990): *Ondelettes et Opérateurs I: Ondelettes*, Hermann Édit.

C. A. Micchelli (1979): On a numerically efficient method for computing multivariate B-splines, in *Multivariate Approximation Theory*, W. Schempp and K. Zeller (eds.), Birkhäuser, Basel, 211–248.

C. A. Micchelli (1980b): A constructive approach to Kergin interpolation in $\mathbb{R}^k$: multivariate B-splines and Lagrange interpolation, Rocky Mountain J. Math. **10**, 485–497.

C. A. Micchelli (1984b): Recent progress in multivariate splines, in *Proceedings of the International Congress of Mathematicians, Vol. 1, 2 (Warsaw, 1983)*, PWN, Warsaw, 1523–1524.

C. A. Micchelli (1986e): Subdivision algorithms for curves and surfaces, in *Extension of B-spline curve algorithms to surfaces*, SIGGRAPH Course #5, C. de Boor (organizer), ACM SIGGRAPH 86, Dallas TX, .

H. G. ter Morsche (1987): Attenuation factors and multivariate periodic spline interpolation, in *Topics in Multivariate Approximation*, C. K. Chui,

L. L. Schumaker, and F. Utreras (eds.), Academic Press, New York, 165–174.

Edward Neuman (1989): Computation of inner products of some multivariate splines, in *Splines in Numerical Analysis (Weissig, 1989)*, Jochen W. Schmidt and Helmuth Spath (eds.), Akademie-Verlag, Berlin, 97–110.

M. Newman (1972): *Integral Matrices*, Academic Press, New York.

M. J. D. Powell (1974): Piecewise quadratic surface fitting for contour plotting, in *Software for Numerical Mathematics*, D. J. Evans (ed.), Academic Press, London, 253–271.

M. J. D. Powell, and M. A. Sabin (1977): Piecewise quadratic approximations on triangles, ACM Trans. Math. Software **3**, 316–325.

H. Prautzsch (1983): Unterteilungsalgorithmen für Bézier und B-spline Flächen, M. S. Thesis, Univ. Braunschweig.

H. Prautzsch (1984c): Unterteilungsalgorithmen für multivariate Splines – ein geometrischer Zugang, dissertation, Univ. Braunschweig.

H. Prautzsch (1985): Generalized subdivision and convergence, Comput. Aided Geom. Design **2**, 69–75.

H. Prautzsch (1986): The location of the control points in the case of box splines, IMA J. Numer. Anal. **6**, 43–49.

H. Prautzsch (1988b): The generation of box spline surfaces, ms.

G. de Rham (1947): Un peu de mathematique à propos d'une courbe plane, Elem. Math. **2**, 73–76, 89–97.

S. D. Riemenschneider (1989): Multivariate cardinal interpolation, in *Approximation Theory VI*, C. Chui, L. Schumaker, and J. Ward (eds.), Academic Press, New York, 561–580.

S. D. Riemenschneider, and K. Scherer (1987): Cardinal Hermite interpolation with box splines, Constr. Approx. **3**, 223–238.

S. D. Riemenschneider, and K. Scherer (1991): Cardinal hermite interpolation with box splines II, Numer. Math. **58**, 133–149.

S. D. Riemenschneider, and Zuowei Shen (1991): Box splines, cardinal series, and wavelets, in *Approximation Theory and Functional Analysis*, C.K. Chui (ed.), Academic Press, New York, 133–149.

S. D. Riemenschneider, and Zuowei Shen (1992): Wavelets and prewavelets in low dimensions, J. Approx. Theory **71**, 18–38.

R. F. Riesenfeld (1975): On Chaikin's algorithm, Computer Graphics and Image Processing **4**, 304–310.

A. Ron (1988): Exponential box splines, Constr. Approx. **4**, 357–378.

A. Ron (1989c): A necessary and sufficient condition for the linear independence of the integer translates of a compactly supported distribution, Constr. Approx. **5**, 297–308.

A. Ron (1990c): Factorization theorems of univariate splines on regular grids, Israel J. Math. **70**, 48–68.

A. Ron (1991a): A characterization of the approximation order of multivariate splines, Studia Math. **98(1)**, 73–90.

A. Ron (1991c): On the convolution of a box spline with a compactly supported distribution: the exponential-polynomials in the linear span, J. Approx. Theory **66(3)**, 266–278.

A. Ron (1992a): Remarks on the linear independence of the integer translates of exponential box splines, J. Approx. Theory **71**, 61–66.

A. Ron (1992b): Linear independence of the translates of an exponential box spline, Rocky Mountain J. Math. **22**, 331–351.

A. Ron, and N. Sivakumar (1993): The approximation order of box spline spaces, Proc. Amer. Math. Soc. **117**, 473–482.

W. Rudin (1973): *Functional Analysis*, McGraw-Hill.

M. A. Sabin (1977): The use of piecewise forms for the numerical representation of shape, dissertation, MTA Budapest.

M. A. Sabin (1989): Open questions in the application of multivariate B-splines, in *Mathematical Methods in Computer Aided Geometric Design*, T. Lyche and L. Schumaker (eds.), Academic Press, N. Y., 529–537.

P. Sablonniere (1981a): De l'existence de spline á support borné sur une triangulation équilatérale du plan, Publication ANO-39, U.E.R. d'I.E.E.A.-Informatique, Université de Lille.

P. Sablonnière (1982a): Bases de Bernstein et approximants splines, dissertation, Univ. Lille.

P. Sablonnière (1982b): Interpolation by quadratic splines on triangles and squares, Computers in Industry **3**, 45–52.

P. Sablonnière (1984a): A catalog of B-splines of degree ≤ 10 on a three direction mesh, Rpt. ANO-132, Univ. Lille.

I. J. Schoenberg (1946a): Contributions to the problem of approximation of equidistant data by analytic functions, Part A: On the problem of smoothing of graduation, a first class of analytic approximation, Quart. Appl. Math. **4**, 45–99.

I. J. Schoenberg (1946b): Contributions to the problem of approximation of equidistant data by analytic functions, Part B: On the problem of osculatory interpolation, a second class of analytic approximation formulae, Quart. Appl. Math. **4**, 112–141.

I. J. Schoenberg (1973a): *Cardinal Spline Interpolation*, CBMS, SIAM, Philadelphia.

I. J. Schoenberg (1973d): Notes on spline functions III: On the convergence of the interpolating cardinal splines as their degree tends to infinity, Israel J. Math. **16**, 87–93.

I. J. Schoenberg (1974a): Cardinal interpolation and spline functions VII. The behavior of cardinal spline interpolation as their degree tends to infinity, J. Analyse Math. **XXVII**, 205–229.

G. C. Shepard (1974): Combinatorial properties of associated zonotopes, Canad. J. Math. **XXVI**, 302–321.

N. Sivakumar (1990a): On bivariate cardinal interpolation by shifted splines on a three-direction mesh, J. Approx. Theory **61**, 178–193.

N. Sivakumar (1990b): Studies in box splines, dissertation, Edmonton, Alberta, Canada.

N. Sivakumar (1991a): Concerning the linear dependence of integer translates of exponential box splines, J. Approx. Theory **64**, 95–118.

Th. Skolem (1950): *Diophantische Gleichungen*, Chelsea Pub. Co., New York.

A. Sommerfeld (1904): Eine besondere anschauliche Ableitung des Gaussischen Fehlergesetzes, in *Festschrift LUDWIG BOLTZMANN gewidmet zum 60. Geburtstage*, Verlag von J. A. Barth, Leipzig, 848–859.

R. Stanley (1973): Linear homogeneous diophantine equations and magic labelings of graphs, Duke Math. J. **43**, 511–531.

R. Stanley (1986): *Enumerative Combinatorics, Vol. 1*, Wadsworth, Belmont CA.

J. Stöckler (1988a): Interpolation mit mehrdimensionalen Bernoulli-Splines und periodischen Box-Splines, dissertation, Duisburg, Germany.

J. Stöckler (1989a): Cardinal interpolation with translates of shifted bivariate box-splines, in *Mathematical Methods in Computer Aided Geometric Design*, T. Lyche and L. Schumaker (eds.), Academic Press, N. Y., 583–592.

J. Stöckler (1989b): Minimal properties of periodic box-spline interpolation on a three-direction mesh, in *Multivariate Approximation Theory IV*, C. Chui, W. Schempp, and K. Zeller (eds.), Birkhäuser Verlag, Basel, 329–336.

J. Stöckler (1992): Multivariate Wavelets, in *Wavelets – A Tutorial in Theory and Applications* (C. K. Chui ed.), Academic Press, New York, 325–355.

G. Strang (1971): The finite element method and approximation theory, in *Numerical Solution of Partial Differential Equations II, SYNSPADE 70*, B. Hubbard (ed.), University of Maryland, College Park, 547–583.

G. Strang (1973): Piecewise polynomials and the finite element method, Bull. Amer. Math. Soc. **79**, 1128–1137.

G. Strang (1974): The dimension of piecewise polynomials, and one-sided approximation, in *Numerical Solution of Differential Equations*, G. A. Watson (ed.), Springer, Berlin, 144–152.

G. Strang, and J. Fix (1973a): *An analysis of the finite element method*, Prentice-Hall, Englewood Cliffs, NJ.

G. Strang, and G. Fix (1973b): A Fourier analysis of the finite element variational method, in *Constructive Aspects of Functional Analysis*, G. Geymonat ed., C.I.M.E. II Ciclo 1971, 793–840.

J. Sun (1986): The approach of Fourier transform to multivariate B-splines, Math. Numer. Sin. **8**, 191–199.

J. Wang (1985): Representations of box-splines by truncated powers, Math. Numer. Sin. **7**, 78–89.

Jian Zhong Wang (1986a): On the expansion coefficients in bivariate box splines, Chinese Annals of Mathematics. Series A. Shuxue Niankan. Ji A **7**, 655–665.

Jian Zhong Wang (1986b): Biorthogonal functionals of box-splines, Math. Numer. Sin. **8**, 75–81.

Jian Zhong Wang (1987): On dual basis of bivariate box-spline, Approx. Theory Appl. **3**, 153–163.

Ren Hong Wang (1975a): Structure of multivariate splines and interpolation, Acta Math. Sinica **18**, 95–106.

J. D. Ward (1987): Polynomial reproducing formulas and the commutator of a locally supported spline, in *Topics in Multivariate Approximation*, C.

K. Chui, L. L. Schumaker, and F. Utreras (eds.), Academic Press, New York, 255–263.

Shu-Ling Zhang (1988): The relationship between box splines and multivariate truncated power functions, J. Northwest-Univ. (J. Northwest-Univ. Natural Sciences (Xibei Daxue Xuebao. Ziran Kexue Ban) **18**, 55–57.

Zuo Shun Zhang (1989a): A multivariate cardinal interpolation problem, Chinese Annals of Mathematics. Series A **10**, 581–587.

Zuo Shun Zhang (1989b): A further discussion on dual bases of bivariate box splines, Chinese Journal of Numerical Mathematics and Applications **11**, 50–58.

P. B. Zwart (1973): Multivariate splines with non-degenerate partitions, SIAM J. Numer. Anal. **10**, 665–673.

Index

Applied Mathematical Sciences

(continued from page ii)

52. *Chipot:* Variational Inequalities and Flow in Porous Media.
53. *Majda:* Compressible Fluid Flow and System of Conservation Laws in Several Space Variables.
54. *Wasow:* Linear Turning Point Theory.
55. *Yosida:* Operational Calculus: A Theory of Hyperfunctions.
56. *Chang/Howes:* Nonlinear Singular Perturbation Phenomena: Theory and Applications.
57. *Reinhardt:* Analysis of Approximation Methods for Differential and Integral Equations.
58. *Dwoyer/Hussaini/Voigt (eds):* Theoretical Approaches to Turbulence.
59. *Sanders/Verhulst:* Averaging Methods in Nonlinear Dynamical Systems.
60. *Ghil/Childress:* Topics in Geophysical Dynamics: Atmospheric Dynamics, Dynamo Theory and Climate Dynamics.
61. *Sattinger/Weaver:* Lie Groups and Algebras with Applications to Physics, Geometry, and Mechanics.
62. *LaSalle:* The Stability and Control of Discrete Processes.
63. *Grasman:* Asymptotic Methods of Relaxation Oscillations and Applications.
64. *Hsu:* Cell-to-Cell Mapping: A Method of Global Analysis for Nonlinear Systems.
65. *Rand/Armbruster:* Perturbation Methods, Bifurcation Theory and Computer Algebra.
66. *Hlavácek/Haslinger/Necasl/Lovísek:* Solution of Variational Inequalities in Mechanics.
67. *Cercignani:* The Boltzmann Equation and Its Applications.
68. *Temam:* Infinite Dimensional Dynamical Systems in Mechanics and Physics.
69. *Golubitsky/Stewart/Schaeffer:* Singularities and Groups in Bifurcation Theory, Vol. II.
70. *Constantin/Foias/Nicolaenko/Temam:* Integral Manifolds and Inertial Manifolds for Dissipative Partial Differential Equations.
71. *Catlin:* Estimation, Control, and the Discrete Kalman Filter.
72. *Lochak/Meunier:* Multiphase Averaging for Classical Systems.
73. *Wiggins:* Global Bifurcations and Chaos.
74. *Mawhin/Willem:* Critical Point Theory and Hamiltonian Systems.
75. *Abraham/Marsden/Ratiu:* Manifolds, Tensor Analysis, and Applications, 2nd ed.
76. *Lagerstrom:* Matched Asymptotic Expansions: Ideas and Techniques.
77. *Aldous:* Probability Approximations via the Poisson Clumping Heuristic.
78. *Dacorogna:* Direct Methods in the Calculus of Variations.
79. *Hernández-Lerma:* Adaptive Markov Processes.
80. *Lawden:* Elliptic Functions and Applications.
81. *Bluman/Kumei:* Symmetries and Differential Equations.
82. *Kress:* Linear Integral Equations.
83. *Bebernes/Eberly:* Mathematical Problems from Combustion Theory.
84. *Joseph:* Fluid Dynamics of Viscoelastic Fluids.
85. *Yang:* Wave Packets and Their Bifurcations in Geophysical Fluid Dynamics.
86. *Dendrinos/Sonis:* Chaos and Socio-Spatial Dynamics.
87. *Weder:* Spectral and Scattering Theory for Wave Propagation in Perturbed Stratified Media.
88. *Bogaevski/Povzner:* Algebraic Methods in Nonlinear Perturbation Theory.
89. *O'Malley:* Singular Perturbation Methods for Ordinary Differential Equations.
90. *Meyer/Hall:* Introduction to Hamiltonian Dynamical Systems and the N-body Problem.
91. *Straughan:* The Energy Method, Stability, and Nonlinear Convection.
92. *Naber:* The Geometry of Minkowski Spacetime.
93. *Colton/Kress:* Inverse Acoustic and Electromagnetic Scattering Theory.
94. *Hoppensteadt:* Analysis and Simulation of Chaotic Systems.
95. *Hackbusch:* Iterative Solution of Large Sparse Systems of Equations.
96. *Marchioro/Pulvirenti:* Mathematical Theory of Incompressible Nonviscous Fluids.
97. *Lasota/Mackey:* Chaos, Fractals and Noise: Stochastic Aspects of Dynamics, 2nd ed.
98. *de Boor/Höllig/Riemenschneider:* Box Splines.
99. *Hale/Lunel:* Introduction to Functional Differential Equations.
100. *Arnol'd/Constantin/Feigenbaum/Golubitsky/Joseph/Kadanoff/Kreiss/McKean/Marsden/Teman:* Trends and Perspectives in Applied Mathematics.
101. *Nusse/Yorke:* Dynamics: Numerical Explorations.
102. *Chossat/Iooss:* The Couette-Taylor Problem.